# Lecture Notes in Mathematics

Edited by A. Dold and B. Eckmann

Subseries: Mathematisches Institut der Universität und
Max-Planck-Institut für Mathematik, Bonn – vol. 8

Adviser: F. Hirzebruch

1231

Ernst-Ulrich Gekeler

# Drinfeld Modular Curves

Springer-Verlag
Berlin Heidelberg New York London Paris Tokyo

**Author**

Ernst-Ulrich Gekeler
Max-Planck Institut für Mathematik
Gottfried-Claren-Str. 26, 5300 Bonn 3, Federal Republic of Germany

Mathematics Subject Classification (1980): 12A90, 10D12, 10D07, 14H25

ISBN 3-540-17201-7 Springer-Verlag Berlin Heidelberg New York
ISBN 0-387-17201-7 Springer-Verlag New York Berlin Heidelberg



Printing and binding: Druckhaus Beltz, Hemsbach/Bergstr.
2146/3140-543210

## Table of Contents

Introduction

0 Notations 1

I Drinfeld Modules

1. Algebraic Theory 2
2. Analytic Theory 5
3. The Operation of $GL(r,A_f)$ 7
4. The Modular Schemes for $r = 1$ and $2$ 9

II Lattices

1. Adelic Description of Lattices 10
2. Lattice Invariants 13
3. Morphisms of Lattices 15

III Partial Zeta Functions

1. Relations with Lattice Sums 17
2. The Rational Function $Z_{a,n}(S)$ 20
3. Evaluation at $s = 0$ and $s = -1$ 22

IV Drinfeld Modules of Rank 1

1. The Case of a Rational Function Field 25
2. Normalization 26
3. Some Lemmata 30
4. Computation of Lattice Invariants 33
5. Distinguished 1-D-Modules 38

V Modular Curves over C

1. The "Upper Half-Plane" 40
2. Group Actions 43
3. Modular Forms 47
4. Elliptic Points 50
5. Modular Forms and Differentials 51

Appendix: The First Betti Number of $\Gamma$ 54

VI Expansions around Cusps

1. Preparations 58
2. Formulae 60
3. Computation of the Factors 61
4. The Δ-Functions 65
5. Some Consequences 71

VII Modular Forms and Functions

1. The Field of Modular Functions 78
2. The Field of Definition of the Elliptic Points 82
3. Behavior of $E^{(q-1)}$ at Elliptic Points 83
4. The Graded Algebra of Modular Forms 85
5. Higher Modular Curves 86
6. Modular Forms for Congruence Subgroups 92

VIII Complements

1. Hecke Operators 94
2. Connections with the Classification of Elliptic Curves 96
3. Some Open Questions 99

Index 101

List of Symbols 102

Bibliography 104

## Introduction

The analogy of the arithmetic of number fields with that of "function fields" (i.e. function fields in one variable over a finite field of constants) has been known for a long time. This analogy starts with elementary things (structure of rings of integers, ramification theory, product formula...), but reaches into such deep fields like for example

- (abelian and non-abelian) class field theory;
- Iwasawa theory;
- special values of L-functions (conjectures of Birch and Swinnerton-Dyer and of Stark, relations with K-theory);
- diophantine geometry (conjecture of Taniyama-Weil).

Many problems in number theory have parallels for function fields; conversely, it is often possible to transfer techniques and geometric considerations from the theory of function fields to the case of number fields.

Within the classical theory of modular forms on the complex upper half-plane and the various generalizations of that theory, one can distinguish between two different points of view:

a) Langlands' philosophy. Here one looks for general reciprocity laws that relate l-adic Galois representations with representations of adele-valued reductive groups.

b) The classical function theoretic approach. Here one is interested in properties of single modular forms (Fourier coefficients, algebraicity, integrality, congruence properties, associated L-series...).

Needless to say that it is neither possible nor reasonable to strictly separate these approaches.

While the main tools of a) come from representation theory and functional analysis, in b) methods of function theory and algebraic geometry are dominating.
As is well known, the representation theoretic approach in its adelic formulation may be transferred to function fields (see e.g. [32,34,40]). In the important paper [11], Drinfeld has shown how to transfer b) as well, i.e. how to obtain a modular theory in the function field case.

Let now $K$ be a function field over the finite field $\mathbb{F}_q$ with $q$ elements, "$\infty$" a fixed place of $K$ of degree $\delta \geq 1$, $A$ the ring of functions in $K$ with poles at most at $\infty$, $K_\infty$ the completion at $\infty$, $C = \hat{\bar{K}}_\infty$ the completion of an algebraic closure of $K_\infty$.

The group $\Gamma = GL(2,A)$ operates by fractional linear transformations on the "upper half-plane" $\Omega = C - K_\infty$, and the set of similarity classes of two-dimensional free discrete A-lattices in $C$ is naturally parametrized by $\Gamma\setminus\Omega$. To each such lattice $\Lambda$, one associates an entire function $e_\Lambda : C \to C$ which will play simultaneously the part of the classical lattice functions $\sigma$, $\zeta$, $\wp$ of Weierstraß. By means of $e_\Lambda$, one constructs an algebraic object over $C$ (later on called a Drinfeld module over $C$ of rank two) whose definition makes sense over arbitrary A-schemes. Proceeding this way, one obtains on $\Gamma\setminus\Omega$ (and, more generally, on $\Gamma'\setminus\Omega$ for each congruence subgroup $\Gamma'$ of $\Gamma$) an algebraic structure as a modular scheme which, roughly speaking, has all the properties of a classical modular curve. $\Gamma'\setminus\Omega$ carries different structures. It is

a) a rigid analytic variety of dimension one over $C$;

b) the set of C-valued points of an affine algebraic curve $M_{\Gamma'}$, which is defined over a finite extension of $K$;

c) "fibred over $\Gamma'\setminus\mathfrak{T}$", where $\mathfrak{T}$ denotes the Bruhat-Tits tree of $PGL(2,K_\infty)$.

In considering c), Drinfeld gives (Thm.2 in [11]) an interpretation of the first l-adic cohomology module of $M_{\Gamma'}$ as a space of automorphic forms in the sense of [40]. (One should note that this theory has been generalized by introducing a level structure "at infinity" (see [12]), but this aspect, leading far away from the classical case, will not be pursued further in this work.)
In contrast to the situation over number fields, there exist r-dimensional A-lattices in $C$ for arbitrary natural numbers $r$ (instead of $r = 2$ only). Correspondingly, we have Drinfeld modules of rank $r$, denoted for short by "r-D-modules".

Let us first consider the case $r = 1$. A 1-D-module has a similar meaning for the arithmetic of $K$ as the multiplicative group scheme $G_m$ for the field $\mathbb{Q}$ of rationals. Drinfeld shows [11, Thm.1]: The modular scheme $M^1$ for 1-D-modules with level structure is $\mathrm{Spec}(B)$, where $B$ is the ring of integers in the maximal abelian extension of $K$ which is totally split at the place $\infty$. This represents a simultaneous analogue for both the theorem of Kronecker-Weber and the main theorem of complex multiplication.

For an arbitrary $r$, a r-D-module behaves, roughly speaking, like an "irreducible abelian variety of dimension $r/2$ over $\mathbb{C}$". For example, the A-module of a-division points (where $a$ is a non-constant element of $A$) is free of rank $r$ over $A/a$. By deformation arguments, Drinfeld obtains the nonsingularity of the modular scheme for the module problem

"r-D-modules + a sufficiently rigid structure of level".

The critical step in Drinfeld's proof of his Thm.2 is to construct a "compactification" $\bar{M}^2$ of the modular scheme $M^2$ for 2-D-modules. Although this is done by ad-hoc glueing a certain one-dimensional scheme to $M^2$ (and not by generalizing the module problem as in [8]), the resulting $\bar{M}^2$ still has a weak modular property with respect to degenerate 2-D-modules. One can give a different construction for $\bar{M}^2 \times C$ which has the advantage of being applicable to the higher ranks $r \geq 2$. For this "Satake-type" compactification, as well as for generalizations of the expansions around cusps given in VI, see [73].

Let us now restrict to the case $r = 2$.

Until recently, not very much was known about the geometry and arithmetic of the curves $M_{\Gamma'}$ ($\Gamma' \subset \Gamma = GL(2,A)$ a congruence subgroup), with the exception of the case $A = \mathbb{F}_q[T]$, the polynomial ring in one indeterminate $T$. In this latter case, the modular scheme $M^2(1) \times C$ for 2-D-modules without structure of level has the curve $M_\Gamma = \Gamma \backslash \Omega$ as its only irreducible component. This curve has genus 0, and is identified with $C$ by means of a j-invariant

$$j : \Gamma \backslash \Omega \xrightarrow{\cong} C .$$

One can calculate the genus of the higher modular curves by the Hurwitz formula. Some other properties of higher modular curves in this special case may be found in [20,21].
Let now again $A$ be arbitrary. One reason for the interest in the curves $M_{\Gamma'}$ concerns diophantine geometry (by a theorem of Grothendieck and Deligne, the analogue of Taniyama-Weil's conjecture on the parametrization

of elliptic curves is true in this context); another one comes from the relations with the cohomology of $\Gamma'$ and with vector bundles over the nonsingular model of $K$ [61]. In view of the work of Ribet, Wiles, Kubert-Lang, and Mazur-Wiles, one should also study the groups of divisor classes of degree zero supported by the cusps of such curves. These groups are finite by (VI 5.12).

Of course, the main tools in the investigation of these curves are modular forms, i.e. C-valued functions on $\Omega$ with the usual transformation behavior and certain holomorphy conditions. The basic examples of modular forms are the Eisenstein series introduced by D.Goss [27], and certain coefficient functions constructed from 2-D-modules varying with $z$ in $\Omega$ (see V.3).
(Note: "Modular forms" are certain C-valued functions, whereas the term "automorphic form" means some characteristic-0-valued mapping, i.e. both have an a priori completely different meaning.)

In this context, we have different "analytic" theories:

a) the theory of the complex zeta function and L-series of $K$. This is well known and presents no analytic difficulties, these functions being rational in $S = q^{-s}$. Nevertheless, the special values of partial zeta functions at the negative integers $1 - r$ $(r \geq 1)$ are of arithmetic interest;

b) the theory of automorphic forms;

c) the C-valued theory of the functions $e_\Lambda$, the modular forms, and related functions.

One link between a) and c) is given by Deligne-Tate's theorem on Stark's conjecture in the function field case [66, Ch.V] resp. by (IV 4.10, 4.13) and (VI 3.9, 4.11). It depends on the distribution property of division points of Drinfeld modules.
(After having a Satake type compactification of the higher rank modular schemes at our disposal, we can generalize this relation: In complete analogy with (IV 4.10) and (VI 4.11), the product expansions of modular forms of rank $r \geq 3$ around cuspidal divisors correspond to the values of partial zeta functions at $1 - r$ [73]. It is not clear, where one should look for a corresponding result in the number field case. Is it reasonable to expect analogous properties for Siegel modular forms?)

Further, modular forms may be considered as multi-differentials on modular curves, whereas automorphic forms occur in the l-adic cohomology of such curves, thereby connecting b) and c).

In this work, modular forms of the Drinfeld type are investigated, and consequences for modular curves are derived. Therefore, representation theory plays no role here. Beyond that, I refrained from discussing Drinfeld's Thm.2, though it certainly gave the motivation for introducing the general theory. For better orientation of the reader, I tried, wherever possible, to indicate the analogies with the number field case.

The emphasis is in the analytic theory of modular forms above $C$ :

- description of $C$-valued points of modular curves;
- behavior of modular forms at cusps: product expansion , properties of the coefficients, zero orders at cusps;
- behavior at elliptic points;
- determination of the $C$-algebra of modular forms....;
- arithmetic consequences.

The relations with algebraic modular forms in the sense of [27] are carried out only as far as possible without going beyond this framework.

One important prerequisite is Hayes' normalization of Drinfeld modules of rank 1. With its help, we may define the generalized cyclotomic polynomials occuring in the product expansions.
As a result, I am able to compute the genera of the modular curves which were not known before (except in the special case mentioned). In particular, an answer is given to the question for the first Betti number of $\Gamma$ (or of groups commensurable with $\Gamma$ ) which has been left open in [61].

The state of problem differs from that in the classical case. For the group $SL(2,\mathbb{Z})$ , one has the well known fundamental domain which leads to the value 0 for the genus of the modular curve "without level structure", and by means of Hurwitz's formula, it is easy to compute the genera of arbitrary modular curves. For $A = \mathbb{F}_q[T]$ , one may follow the same lines, this case being treated in [17]. However, for a general $A$ , the equations defining Drinfeld modules become so complicated that, already in the next simple case (genus of $K = 0$ and $\delta = 2$ ), it seems hopeless to try to compute the genera in this naive way. Using the Bruhat-

Tits tree, it is possible to construct a fundamental domain for $\Gamma$ in some very restricted cases [61], but this does not help much. Instead, we use the description of the elliptic points and of the parameters at cusps to get a relation between modular forms for $\Gamma$ and multi-differentials on $\bar{M}_\Gamma$. This allows the computation of $g(\bar{M}_\Gamma)$, if the divisor of one single modular form is known. Finally, the divisors of certain modular forms are obtained from the above mentioned product expansions.

The organization of the work is as follows:

Since Drinfeld modules do not (as yet) belong to the basic tools of the number theorist, the needed definitions, concepts and properties are collected in Chapter I. It contains no proofs; these may be found in [10,11] and, partially, in [36]. The deepest facts cited are the non-singularity of the modular schemes (1.10), the description (4.1) of $M^1$ and the compactification of $M^2$ (4.2).

In Chapter II, one finds properties of lattices often needed throughout the work (§ 1), relations between the coefficients of power series associated with lattices, and the links with lattice sums (§2), as well as additive polynomials related to morphisms of lattices (§ 3). A simple, but very important fact is (2.10): lattices $\Lambda$ resp. Drinfeld modules are completely determined by the values of finitely many Eisenstein series $E^{(k)}(\Lambda)$.

Still some preparations are done in Chapter III. In the first paragraph, well known facts on the zeta functions of $K$ and $A$ are collected, and partial zeta functions for elements of Pic $A$ are defined and compared with complex valued lattice sums. We obtain a distribution on the set of pairs $(a,\mathfrak{n})$, where $a$ lies in $K$ and $\mathfrak{n}$ is a fractional A-ideal in $K$. This distribution takes values in the field $\mathbb{C}(S)$ of rational functions $(S = q^{-s})$. Its evaluation at places $s = 1-r$ describes a $\mathbb{C}$-valued distribution which, later on, will turn out to be the distribution constructed from division points of Drinfeld modules of rank r. An explicit presentation of the rational function $Z_{a,\mathfrak{n}}$ by means of generalized Weierstraß gaps is given in § 2. In the third section, certain finite sums occurring later are interpreted as values of $Z_{a,\mathfrak{n}}$ at $S = 1$ resp. $q$, i.e. as zeta values at $s = 0$ resp. $s = -1$. There is a uniform upper bound for the numbers $Z'_{a,\mathfrak{n}}(1)$ which assures the convergence of our product expansions in VI.

In Chapter IV, we deal with Drinfeld modules of rank 1 over $C$. This will be necessary for the rank 2 theory, but is also interesting for its own sake. First, we handle the most simple case $A = \mathbb{F}_q[T]$ . Here, all 1-D-modules are isomorphic with the module studied by Carlitz [4,5,6]. Its division points generate the maximal abelian extension of $K$ which is completely split at $\infty$ . The analogy with the Kronecker-Weber theorem is obvious, so this example will serve as a motivation for what follows. For arbitrary $A$ , the isomorphism classes of 1-D-modules are parametrized by Pic $A$ . For generalizing the above example, we need "canonical" 1-D-modules, i.e. for each isomorphism class a distinguished module. These modules are not given by Drinfeld's theory. If $\delta = 1$ , it is easy to see:

> For each element of Pic $A$ , there exists a 1-D-module (uniquely determined up to trivial transformations) with coefficients in the ring of integers of the ring class field $H$ of $A$ and leading coefficients in $\mathbb{F}_q^*$ .

Under this assumption, the wanted generalization causes no problems [37]. This is no longer true for $\delta > 1$ . In [39], Hayes shows how to proceed in the general case to generate class fields of $K$ by division points of D-modules. First, one has to choose a sign function sgn , i.e. a co-section of the embedding $\mu_w \hookrightarrow K_\infty^*$ ,where $w = q^\delta - 1$ . Then one considers D-modules $\phi$ with the following property: The function $A \to C$ which associates to each $a \in A$ the leading coefficient of the additive polynomial $\phi_a$ agrees up to Galois twist with $\mathrm{sgn}|A$ . In each isomorphism class, there exist such $\phi$ . They are uniquely determined up to twists with $w$-th roots of unity, and have coefficients in a finite abelian extension $\tilde{H}$ of $K$ which contains $H$ . In § 2, we give, as far as needed, and without proofs, a summary of Hayes' theory of "sgn-normalization".

Now we are able to define the $\xi$-invariants of rank 1 lattices up to $w$-th roots of unity. In sections 3 and 4, these invariants are computed. We obtain product formulae (4.10, 4.13) analogous with the classical

$$\pi = 2 \prod_{a \geq 1} (1 - 1/4a^2)^{-1} .$$

Perhaps the most striking consequence is the relation of such formulae with the values of derivatives of partial zeta functions at $s = 0$ . Proceeding, one can construct units in abelian extensions of $K$ with absolute values (at the different infinite places) prescribed by Stark's conjectures. This gives in fact a constructive proof of Deligne-Tate's theorem on Stark's conjecture in our situation (see [39]). Another result is the determination of the Galois twist by which the "leading coefficient function" differs from sgn (4.11). For later computations in the rank 2 case, we have to fix our $\xi$-invariants. By means of (4.11), we have control on the effect of the choices made, and we are able to describe the isogenies of different D-modules.

In V, we come to the central point. First, Drinfeld's upper half-plane is described in more detail (building mapping, analytic structure). In § 2, we show how the analytic space $\Gamma\backslash\Omega$ is compactified by adjoining a finite number of cusps. In the next section, modular forms are introduced, their behavior at cusps is discussed and some examples are given for the construction of forms by means of 2-D-modules. Elliptic points of the groups $GL(Y)$ are investigated in section 4 (existence, number, structure of stabilizers). This relates modular forms and differentials. The resulting formula (5.5) expresses the genus of a modular curve in terms of the divisor of a modular form. The chapter ends with an appendix not further used in this work. Up to some details, which may be found in [11], a proof of $g(\bar{M}_{\Gamma'}) = b(\Gamma')$ is given. With the results of V and VI, one obtains the first Betti number $b(\Gamma')$ for all arithmetic subgroups $\Gamma' \subset \Gamma$ (not only for those which are p'-torsion free [61]).

Chapter VI is devoted to the computation of expansions of modular forms around cusps. After some preliminaries, in § 3, the expansions of the division functions $e_u$ (some sort of Fricke functions) around the cusp $\infty$ are determined. A major ingredient is the rank 1 theory developed in IV. The result is (3.9) which presents $e_u$ as an infinite product with positive radius of convergence. The pole order of $e_u$ can be expressed, in view of (III 3.11), by zeta values. The fourth section uses these results for the computation of similar product expansions for the "discriminant functions" $\Delta_{\mathfrak{n}}$ associated with positive divisors $\mathfrak{n}$. For principal divisors $\mathfrak{n} = (f)$, this product takes the particularly simple form (4.12), which is, on the one hand, a translation of

$$\Delta = (2\pi i)^{12}\, q \prod_{n \geq 1} (1-q^n)^{24} \;;$$

on the other hand, it is a two-dimensional analogue of the products for the $\xi$-invariants in IV. The determination of the root-of-unity factor in (4.12) is somewhat delicate, because some of the preceding calculations yield results only up to (q-1)-st roots of unity.

The transfer to other cusps is easy. This is carried out in § 5, where we also draw some conclusions:

- final determination of the genus for the modular curves associated to maximal arithmetic subgroups;
- existence of a distinguished cusp form of weight $q^{2\delta} - 1$ ;
- finiteness of the group of cuspidial divisor classes of degree 0.

In VII, the results of VI are transferred to higher modular curves, and rings of modular forms are computed. First, the function field of the modular scheme $M^2(\mathfrak{n})$ is determined, as well as the field of definition of the cusps and of the elliptic points. After the preceding considerations, it suffices to apply some well known arguments (see for example [62]). For being able to compute the dimensions of the spaces of modular forms, it is (up to a small number of exceptions) enough to know the behavior of the Eisenstein series of weight $q-1$ at the elliptic points. In § 3, these series are shown to have simple zeroes at elliptic points. So we are able to give, for the present, dim $M_k(\Gamma)$ for maximal arithmetic groups. Nevertheless, the arithmetic meaning of the modular forms occurring is not at all clear already in the simple examples discussed in § 4. It would be desirable to have a description by generators and relations, where the generating modular forms should have an elementary interpretation by means of Drinfeld modules. In § 5, the genera of modular curves for full congruence subgroups $\Gamma(\mathfrak{n})$ are computed, and a formula is given for the Hecke congruence subgroup $\Gamma_0(\mathfrak{n})$ , in the case $\mathfrak{n} = \mathfrak{p}$ is a prime ideal. (If one works patiently enough, it is possible and not too difficult, to write down a generally valid formula. The corresponding Betti number depends only on the decomposition type of the divisor $\mathfrak{n}$ and, of course, the zeta function of $K$ .) Finally, the dimensions of $M_k(\Gamma')$ are given for some congruence subgroups $\Gamma'$ of $\Gamma = GL(2,A)$ . For $k = 1$ , we get only a lower bound for the dimension.

The final chapter VIII contains some additional material and remarks on relations with other questions. In § 1, the Hecke operators $T_{\mathfrak{p}}$ are introduced. A priori, $T_{\mathfrak{p}}$ is a correspondence on the set of 2-lattices in $C$ . One obtains

a) a correspondence on the modular scheme $\bar{M}^2(\mathfrak{n})$ ;

b) an operator on characteristic-zero valued automorphic forms;

c) an operator on modular forms.

The Eisenstein series are easily seen to be eigenvectors for the $T_{\mathfrak{p}}$ ( $\mathfrak{p}$ = principal ideal), whilst the effect of Hecke for instance on the discriminant functions $\Delta_{\mathfrak{n}}$ is totally unknown. In the second section, the connection with the classification of elliptic curves is discussed. Finally, some questions are raised which have or have not a counterpart in the number field case.

For ease of handling, we have included an index and a list of symbols. References inside the text are made in the form

(x.y) number x.y in the present chapter;

(V x.y) number x.y in Chapter V;

[xy] item xy in the Bibliography.

The end of a proof is labelled by □ . The symbols $\mathbb{N}$, $\mathbb{Z}$, $\mathbb{Q}$, $\mathbb{R}$, $\mathbb{C}$ denote the usual number sets.

$\#(S)$ is the cardinality of a set $S$, $X - Y$ the complement of $Y$ in $X$, $f|Y$ the restriction of the map $f$ to $Y$.

For a ring $R$ and $r$ in $R$, $R^*$, $(r)$, $R/r$ denote the multiplicative group, the principal ideal generated by $r$, the factor ring respectively. The group $G$ acting on $X$, $G\backslash X$ resp. $G/X$ is the orbit space, $X^G$ the fixed point set and $G_x$ the stabilizer of $x \in X$. For $g,h$ in $G$, $h^g = ghg^{-1}$. Further, $\mathrm{Gal}(L:K)$ is the Galois group of the field extension $L:K$, $\bar{K}$ an algebraic closure of $K$ and $\mu_n$ the group of $n$-th roots of unity.

"RS" is the abbreviation for "system of representatives",

"oBdA" means "without loss of generality",

and "$N >> 0$" says "the number $N$ is sufficiently large".

The present text is a slightly complemented english translation of the authors "Habilitationsschrift" at the Faculty of Sciences, Bonn 1985. He wants to thank the staff of the "Max-Planck-Institut für Mathematik" in Bonn for support. In particular, he is grateful to Miss M.Grau who did an excellent job in preparing the manuscript.

## 0. Notations

Throughout the text, $q$ denotes a power of the prime number $p$, and $\mathbb{F}_q$ is the finite field with $q$ elements. Let further

$K$ be a function field in one variable over the field of constants $\mathbb{F}_q$, of genus $g$;

$\infty$ a place of $K$ fixed once for all, of degree $\delta \geqq 1$;

$A$ the ring of functions $f$ in $K$ with poles at most at $\infty$;

$K_\infty$ the $\infty$-adic completion of $K$, with ring of integers $O_\infty$ and residue field $k$.

We choose a uniformizing parameter $\pi$ at $\infty$, and we determine the degree function deg and the absolute value on $K_\infty$ by

$$\deg \pi = -\delta , \ |x| = q^{\deg x} .$$

In particular, $\deg 0 = -\infty$. Divisors on $K$ which are prime to $\infty$ are written multiplicatively and identified with fractional ideals of $A$. They are denoted by $\mathfrak{a},\mathfrak{b}...\mathfrak{m},\mathfrak{n}$. Correspondingly, $\mathfrak{p},\mathfrak{q}...$ are places resp. prime ideals of $A$.

For a divisor $\mathfrak{a}$, let $|\mathfrak{a}| = q^{\deg \mathfrak{a}}$ and $\mathfrak{a}_N = \{a \in \mathfrak{a} | \deg a \leq N\}$. We use "$\mathfrak{a} > 1$" or "$\mathfrak{a} \subset A$" to designate positive divisors $\mathfrak{a}$. Further, we need

$\mathbb{A} = \mathbb{A}_f \times K_\infty$ the ring of adeles of $K$, with finite part $\mathbb{A}_f$; correspondingly, we let

$I = I_f \times K_\infty^*$ the group of ideles of $K$ and

$E = E_f \times E_\infty$ the group of unit ideles.

For $\mathfrak{a} > 1$, let $E(\mathfrak{a}) = E_f(\mathfrak{a}) \times E_\infty = \{\underline{e} \in E | \underline{e} \equiv 1 \bmod \mathfrak{a}\}$. Finally,

$\hat{A} = \varprojlim_{\mathfrak{n}>1} A/\mathfrak{n}$ is the ring of integral finite adeles. As occasion demands, we consider $K$ as a subring of $\mathbb{A}, \mathbb{A}_f$, or $K_\infty$.

# I. Drinfeld Modules

## 1. Algebraic Theory [10,11,36]

(1.1) Let $L$ be a field of characteristic $p$ and $\mathrm{End}_L(G_a)$ the ring of those endomorphisms of the additive group scheme $G_a$ which are defined over $L$. Then $\mathrm{End}_L(G_a)$ is a non-commutative polynomial ring over $L$, generated by the Frobenius endomorphism

$$\tau_p : L \longrightarrow L$$
$$x \longmapsto x^p .$$

We write $\mathrm{End}_L(G_a) = L\{\tau_p\}$, the curly braces indicating the commutation rule $\tau_p x = x^p \tau_p$ for $x \in L$. By $\tau_p \longmapsto X^p$, $\mathrm{End}_L(G_a)$ is isomorphic with the ring of additive polynomials over $L$, i.e. the ring of polynomials of the form

$$\sum l_i X^{p^i} ,$$

the multiplication being defined by substitution. We do not distinguish between both points of view, and we write "$\tau_p^i$" or "$X^{p^i}$", depending on the context.

The structure of $\mathrm{End}_L(G_a)$ has first been studied by Ore [53]; for example, $\mathrm{End}_L(G_a)$ is right euclidean, and each left ideal is principal.

(1.2) We now assume that $L$ has a structure $\gamma : A \to L$ as an A-algebra. By definition, the <u>characteristic of $L$</u> is the prime ideal $\infty$, if $\gamma$ is injective, and $\mathrm{Ker}\,\gamma$ otherwise. An injective ring homomorphism

$$\phi : A \longrightarrow \mathrm{End}_L(G_a)$$
$$a \longmapsto \phi_a$$

defines by

$$\|a\| = \text{degree of the additive polynomial corresponding to } \phi_a$$

an absolute value $\| \ \|$ on $A$, provided there exists an $a$ with $\|a\| > 1$. Under this assumption, the extension of $\| \ \|$ to $K$ is equivalent with $| \ |$. Hence, there exists a real number $r > 0$ such that for all $a$ in $K$, we have $\|a\| = |a|^r$. In fact, $r$ is even a natural number, and $\phi$ takes values in $L\{\tau\} \subset \mathrm{End}_L(G_a)$. Here, $\tau = \tau_p^s$ is the element corresponding to $X^q$, where $q = p^s$. Each element $f$ of $L\{\tau\}$ can be written uniquely in the form $f = \sum l_i \tau^i$ with left coefficients $l_i = l_i(f)$. We put $D(f) = l_o(f)$ = "constant term" of $f$ and $l(f) = l_{\deg f}$ = "leading coefficient" of $f$, where $\deg f$ is the degree of $f$ in $\tau$.

1.3. Definition. A Drinfeld module over $L$ of rank $r \in \mathbb{N}$ is an injective ring homomorphism

$$\phi : A \longrightarrow \mathrm{End}_L(G_a)$$
$$a \longmapsto \phi_a ,$$

such that for all $a \in A$, we have

(i) $\deg \phi_a = r \cdot \deg a$ ($\deg \phi_a$ = degree of $\phi_a$ in $\tau$), and

(ii) $D(\phi_a) = \gamma(a)$.

We abbreviate the notation "Drinfeld module" resp. "Drinfeld module of rank r" by "D-module" resp. "r-D-module".

By $\phi$, the additive group scheme over $L$ becomes a scheme of A-modules. Let $\phi$ and $\Psi$ be D-modules over $L$. A morphism $u : \phi \to \Psi$ is a L-endomorphism $u$ of $G_a$ with the property

$$u \circ \phi_a = \Psi_a \circ u$$

for all $a$ in $A$. If $u$ is an automorphism of $G_a$, i.e. a constant different from 0, $u$ is called an isomorphism. Non-trivial morphisms $u$ are already in $L\{\tau\} \subset L\{\tau_p\}$; they may exist only between Drinfeld modules of the same rank and are therefore called isogenies.

1.4. Example. If $K$ is the field $\mathbb{F}_q(T)$ of rational functions and $A$ the ring $\mathbb{F}_q[T]$ of polynomials in an indeterminate $T$, a r-D-module $\phi$ is given by

$$\phi_T = \sum_{0 \le i \le r} l_i \tau^i ,$$

where $l_i \in L$ , $l_r \neq 0$ , $l_o = \gamma(T)$ .

(1.5) For a D-module $\phi$ and $a$ in $A$ let $D(\phi,a)$ be the A-submodule scheme $\mathrm{Ker}\ \phi_a \subset G_a$ . For an ideal $\mathfrak{n} \subset A$ , let

$$D(\phi,\mathfrak{n}) = \bigcap_{a\in\mathfrak{n}} D(\phi,a)$$

the <u>scheme of $\mathfrak{n}$-division points</u> of $\phi$ .

It is not difficult to show the following

<u>1.6. Proposition</u>. Let $r$ be the rank of the D-module $\phi$ over $L$ .

(i) $D(\phi,\mathfrak{n})$ is a finite group scheme of degree $\#(A/\mathfrak{n})^r$ over $L$ .

(ii) If $\mathfrak{n}$ is prime to the characteristic of $L$ , $D(\phi,\mathfrak{n})(\bar{L})$ is a free $A/\mathfrak{n}$-module of rank $r$ .

Roughly speaking, Drinfeld modules of rank r behave like "irreducible abelian varieties of dimension $r/2$ " in number theory.

We would like to have a universal algebraic family of D-modules. Hence, we are led to extend our definition (1.3) to the case of arbitrary A-schemes $S$ (instead of $S = \mathrm{Spec}\ L$) . Let $\gamma^* : S \to \mathrm{Spec}\ A$ be an A-scheme dual to the ring homomorphism $\gamma : A \to \mathcal{O}_S$ .

<u>1.7. Definition</u>. Let $\phi : A \longrightarrow \mathrm{End}_S(\mathcal{L})$

$$a \longmapsto \phi_a$$

be a homomorphism of $A$ into the ring of endomorphisms of a line bundle ( = locally free sheaf of rank 1) $\mathcal{L}$ over $S$ . $\phi$ is called a Drinfeld module over $S$ of rank r, provided that, for each $a$ in $A$ , we have

(i) locally, as a polynomial in $\tau$, $\phi_a$ has the degree $r \cdot \deg a$ and a unit as its leading coefficient;

(ii) $l_o(\phi_a) = \gamma(a)$ .

Correspondingly, one defines morphisms of D-modules as well as the schemes of division points over $S$ . (I refrain to formally write down these definitions.)

(1.8) The subscheme $D(\phi,\mathfrak{n}) \hookrightarrow \mathcal{L}$ of $\mathfrak{n}$-division points of the r-D-module $\phi$ over $S$ is flat and finite over $S$ and étale outside of the support $\mathrm{supp}(\mathfrak{n})$ of $\mathfrak{n} \subset A$ . From a naive point of view, a level

$\mathfrak{n}$ structure on $\phi$ should be an isomorphism of the constant $(A/\mathfrak{n})$ - module scheme $(\mathfrak{n}^{-1}/A)^r$ with $D(\phi,\mathfrak{n})$ . In order to handle the ramified places too, one uses the

1.9. Definition. A level $\mathfrak{n}$ structure on $\phi$ is a morphism $\alpha : (\mathfrak{n}^{-1}/A)^r \to D(\phi,\mathfrak{n})$ of A-module schemes such that on $\mathcal{L}$ , the identity

$$\sum_{n \in (\mathfrak{n}^{-1}/A)^r} \alpha(n) = D(\phi,\mathfrak{n})$$

of divisors holds. (An intensive discussion of level structures of this type may be found in [42].)

In general, the functor, the functor $r - \mathfrak{D}\mathfrak{M}$

$$S \longmapsto \left\{ \begin{array}{l} \text{isomorphism classes of Drinfeld} \\ \text{modules of rank } r \text{ over } S \end{array} \right\}$$

on the category of A-schemes is not representable. However, if we consider instead the functor $r - \mathfrak{D}\mathfrak{M}(\mathfrak{n})$

$$S \longmapsto \left\{ \begin{array}{l} \text{isomorphism classes of r-D-modules} \\ \text{over } S \text{ with a level } \mathfrak{n} \text{ structure} \end{array} \right\},$$

we have

1.10. Theorem [11, § 5]. If $\mathfrak{n}$ has at least two different prime divisors, $r - \mathfrak{D}\mathfrak{M}(\mathfrak{n})$ is representable by an affine A-scheme $M^r(\mathfrak{n})$ of finite type. $M^r(\mathfrak{n})$ is smooth of dimension $r$ as a scheme over $\mathbb{F}_q$ , and the structural morphism $M^r(\mathfrak{n}) \to \operatorname{Spec} A$ is smooth outside of the support $\operatorname{supp}(\mathfrak{n})$ .

A positive divisor $\mathfrak{n}$ is called admissible, if $\#(\operatorname{supp}(\mathfrak{n})) \geq 2$ .

## 2. Analytic Theory [10,11]

The absolute value on $K_\infty$ has a unique extension to the algebraic closure $\bar{K}_\infty$ , which will also be denoted by "$| \ |$" . The field $\bar{K}_\infty$ is not complete; however, the completion $C$ of $\bar{K}_\infty$ is again algebraically closed [2, Prop.3, p.146]. For such fields, there is a highly developed function theory, see the bibliography in [2].

<u>2.1. Definition.</u> (i) An A-lattice in $C$ is a finitely generated A-submodule $\Lambda$ of $C$ which has a finite intersection with each ball in $C$ of finite radius. $\Lambda$ is called an "r-lattice" if it is projective of rank r. Lattices which agree up to a scalar factor $c \in C^*$ are called <u>similar</u>.

(ii) The exponential function (or lattice function) $e_\Lambda$ of a lattice $\Lambda$ is the infinite product

$$e_\Lambda(t) = t \prod_{\lambda \in \Lambda}{}' (1-t/\lambda) .$$

Here, and in the sequel, we have the usual

<u>Convention</u>. $\prod'$ resp. $\sum'$ are products resp. sums over the non-zero elements of a lattice.

It is easy to show the following properties of $e_\Lambda$ :

(2.2) (i) The product converges, uniformly on bounded sets, and defines an entire function $e_\Lambda : C \to C$ .

(ii) $e_\Lambda$ has simple zeroes at the points in $\Lambda$ and no further zeroes, and, up to constant multiples, it is unique with these properties.

(iii) $e_\Lambda$ is $\mathbb{F}_q$-linear and surjective.

(iv) The functions of similar lattices are related by

$$e_{c\Lambda}(ct) = ce_\Lambda(t) .$$

(v) The derivative $e'_\Lambda$ equals the constant 1. Hence we have the identity of meromorphic functions on $C$ :

$$e_\Lambda^{-1}(t) = e'_\Lambda(t)/e_\Lambda(t) = \sum_{\lambda \in \Lambda} 1/(t-\lambda) .$$

As in complex function theory (where, by means of the Weierstraß function, one associates an elliptic curve with a lattice in $\mathbb{C}$ ), we use the function $e_\Lambda$ to construct a Drinfeld module.

So, let a r-lattice $\Lambda$ be given, and let $\phi_a^\Lambda \in \mathrm{End}_C(G_a)$ be determined by the commutative diagram with exact rows:

(2.3)
$$\begin{array}{ccccccccc} 0 & \longrightarrow & \Lambda & \longrightarrow & C & \xrightarrow{e_\Lambda} & C & \longrightarrow & 0 \\ && \downarrow a && \downarrow a && \downarrow \phi_a^\Lambda && \\ 0 & \longrightarrow & \Lambda & \longrightarrow & C & \xrightarrow{e_\Lambda} & C & \longrightarrow & 0 \; . \end{array}$$

Then $a \longmapsto \phi_a^\Lambda$ is a ring homomorphism $\phi^\Lambda : A \to \mathrm{End}_C(G_a)$ . In fact, $\phi^\Lambda$ is a Drinfeld module of rank r which fully determines the lattice $\Lambda$ (compare (II 2.5)). One obtains

2.4. Theorem. The association $\Lambda \longmapsto \phi^\Lambda$ defines a bijection of the set of r-lattices in $C$ with the set of r-D-modules over $C$ .

2.5. Remark. (i) If one defines morphisms of lattices as numbers $c \in C$ with $c\Lambda \subset \Lambda'$ , (2.4) gives an equivalence of categories.

(ii) Let $\alpha$ be a level (a) structure on $\phi^\Lambda$ . In the commutative diagram

$$\begin{array}{ccc} (a^{-1}/A)^r & \xrightarrow[\alpha]{\cong} & \mathrm{Ker}\ \phi_a^\Lambda \\ \downarrow a & & \downarrow \text{given by (2.3)} \\ (A/a)^r & \xrightarrow[\beta]{} & \Lambda/a\Lambda \end{array} ,$$

the vertical máps depend on the choice of $a$ , whereas $\beta$ is defined by $\alpha$ and the ideal (a) . This shows, a positive divisor $\mathfrak{n}$ being given, the equivalence of the following data which will not be distinguished in the sequel:

a) level $\mathfrak{n}$ structures on $\phi^\Lambda$ ;

b) isomorphisms of $(\mathfrak{n}^{-1}/A)^r$ with $\mathfrak{n}^{-1}\Lambda/\Lambda$ ;

c) isomorphisms of $(A/\mathfrak{n})^r$ with $\Lambda/\mathfrak{n}\Lambda$ .

## 3. The Operation of $GL(r,\mathbb{A}_f)$ [11, § 5]

Let $G$ be the group scheme $GL(r)$ with center $Z$ .

(3.1) If $\mathfrak{n}$ is an admissible ideal, the modular scheme $M^r(\mathfrak{n})$ is well-defined and affine, and for $\mathfrak{m} \subset \mathfrak{n}$, we have the finite forget morphisms $M^r(\mathfrak{m}) \to M^r(\mathfrak{n})$. Therefore, the projective limit

$$M^r = \varprojlim_{\mathfrak{n} \subset A} M^r(\mathfrak{n})$$

exists.

(3.2) We define an operation of the group $G(\mathbb{A}_f)$ on $M^r$. Note first the equality

$$G(\mathbb{A}_f) = Z(K) \cdot W ,$$

where $W$ is the semigroup of matrices with coefficients in $\hat{A}$. Let $\phi$ be a r-D-module over an A-scheme $S$ with a level structure

$$\alpha : (K/A)^r \to D(\phi) = \bigcup_{a \in A} D(\phi, a) ,$$

i.e. for each ideal $\mathfrak{n}$, $\alpha$ induces a level $\mathfrak{n}$ structure. An element $\underline{g}$ of $W$ acts from the left on $(K/A)^r$. Its kernel is mapped by $\alpha$ onto an A-submodule scheme $U$ of $\phi$. Let $\underline{g}_* : \phi \to \phi'$ be a morphism of D-modules over $S$ with kernel $U$. Define now $\alpha'$ such that the diagram

$$\begin{array}{ccc} (K/A)^r & \xrightarrow{\alpha} & D(\phi) \\ \downarrow \underline{g} & & \downarrow \underline{g}_* \\ (K/A)^r & \xrightarrow{\alpha'} & D(\phi') \end{array}$$

commutes.

Then $\alpha'$ is a level structure on $\phi'$, and $(\phi', \alpha')$ is uniquely determined by $\underline{g}$ up to isomorphism. The operation of $W$ on $M^r$ given by $(\underline{g}, (\phi, \alpha)) \longmapsto (\phi', \alpha')$ has a unique extension to an operation of the group $G(\mathbb{A}_f)$, where $Z(K)$ acts trivially. The corresponding operation of $G(\mathbb{A}_f)/Z(K)$ is effective.

(3.3) We may now use this operation to define modular schemes $M^r(\mathfrak{n})$ even for non-admissible divisors $\mathfrak{n}$. Let $G(\hat{A}, \mathfrak{n})$ be the kernel of the reduction map $G(\hat{A}) \to G(A/\mathfrak{n})$. For admissible $\mathfrak{n}$, $M^r(\mathfrak{n})$ exists and equals $G(\hat{A}, \mathfrak{n}) \backslash M^r$. Hence we define for arbitrary open subgroups $K$ of $G(\hat{A})$

$$M_K^r = K \backslash M^r \quad .$$

In any case, $M_K^r$ will be at least a coarse modular scheme for the module problem corresponding to $K$. If $K$ equals some $G(\hat{A},\mathfrak{n})$, we write $M^r(\mathfrak{n})$ even for non-admissible $\mathfrak{n}$. In particular, for the full group $K = G(\hat{A})$, we obtain the coarse modular scheme $M^r(1)$ associated to the module problem "Drinfeld modules of rank r".

(3.4) The intersection $Z(K) \cap G(\hat{A})$ in $G(A_f)$ is $Z(\mathbb{F}_q)$. Considering $\mathbb{F}_q$ as a subring of $A/\mathfrak{n}$ ( $\mathfrak{n}$ supposed to be a proper ideal of $A$ ), we get

$$G(\hat{A})/G(\hat{A},\mathfrak{n}) \cdot Z(\mathbb{F}_q) = G(A/\mathfrak{n})/Z(\mathbb{F}_q)$$

as the group of the Galois covering $M^r(\mathfrak{n}) \to M^r(1)$.

## 4. The Modular Schemes for $r = 1$ and $2$

By (3.2), $M^1$ is an affine scheme with an effective operation of $I_f/K^*$. In fact, we have

4.1. Theorem [11, Thm.1]: $M^1$ is the spectrum of the integral closure of $A$ in the maximal abelian extension of $K$ which is totally split at $\infty$. Considering $I_f/K^*$ as the quotient $I/K^* \cdot K_\infty^*$ of the idele class group of $K$, the operation on $M^1$ agrees with that of class field theory.

For another approach, compare [36].

For the modular schemes $M^2(\mathfrak{n}) \to \operatorname{Spec} A$, there exist canonical compactifications.

4.2. Theorem [11, Prop.9.3]: Let $\mathfrak{n}$ be admissible.

(i) There exists a uniquely determined smooth two-dimensional A-scheme $\bar{f} : \bar{M}^2(\mathfrak{n}) \to \operatorname{Spec} A$ such that in the commutative diagram

$$\begin{array}{ccccc} M^2(\mathfrak{n}) & \overset{i}{\hookrightarrow} & \bar{M}^2(\mathfrak{n}) & \hookleftarrow & \bar{M}^2(\mathfrak{n}) - M^2(\mathfrak{n}) \\ & {\scriptstyle f}\searrow & \downarrow{\scriptstyle \bar{f}} & \swarrow{\scriptstyle f_c} & \\ & & \operatorname{Spec} A & & \end{array}$$

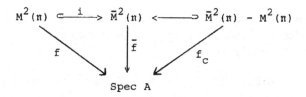

the following assertions hold:

a) $i$ is an open dense imbedding;

b) $f$ is the structural morphism of $M^2(\mathfrak{n})$ ;

c) $\bar{f}$ is proper;

d) $f_C$ is finite.

(ii) The canonical morphisms $M^2(\mathfrak{m}) \to M^2(\mathfrak{n})$ have extensions to finite morphisms $\bar{M}^2(\mathfrak{m}) \to \bar{M}^2(\mathfrak{n})$ , and the action of $G(\mathbb{A}_f)$ can be extended to $\bar{M}^2 = \varprojlim \bar{M}^2(\mathfrak{n})$ .

(iii) $\bar{f} : \bar{M}^2(\mathfrak{n}) \to \mathrm{Spec}\ A$ is smooth outside of $\mathrm{supp}(\mathfrak{n})$ .

As in (3.3), we define $\bar{M}^2_K$ for open subgroups $K$ of $GL(2,\hat{A})$ . An explicit description of the set of C-valued points of $\bar{M}^2_K$ will be given in (V 2).

## II Lattices

### 1. Adelic Description of Lattices

We are giving here some well known facts from lattice theory and use them to describe the modular schemes $M^r(\mathfrak{n})$ . In the whole section, $G$ is the group scheme $GL(r)$ .

(1.1) For a finitely generated A-submodule $Y$ of $K^r$ , the following assertions are equivalent:

(i) $Y \otimes K_\infty \xrightarrow{\cong} K_\infty^r$ ;

(ii) $Y$ is projective of rank $r$ and discrete in $K_\infty^r$ ;

(iii) $Y$ generates $K^r$ and is discrete in $K_\infty^r$ .

Any $Y$ with these properties is called an <u>r-lattice</u>.

(If there might arise soem confusion with the r-lattices of (I 2.1), we will denote the objects considered there by "r-lattices in C ", and those considered here by "lattices in $K^r$ ".)

(1.2) A matrix $\underline{g}$ in $G(\mathbb{A}_f)$ defines in a well known manner a r-lattice $Y = Y(\underline{g})$, characterized by

$$Y \cdot \hat{A} = (\hat{A}^r)g^{-1} \subset A_f^r .$$

(Note: $G(A_f)$ acts from the right as a group of matrices on $A_f^r$ .) Each r-lattice may be constructed this way, e.g. [33, § 2].

(1.3) Two lattices $Y$, $Y'$ are isomorphic if and only if there exists an element $\gamma$ of $G(K)$ with $Y\gamma = Y'$ . This gives a bijection

$$G(K)\backslash G(A_f)/G(\hat{A}) \xrightarrow{\cong} P_A^r ,$$

$$g \longmapsto \text{class of } Y(g)$$

where $P_A^r$ denotes the set of isomorphism classes of r-lattices ( = set of classes of projective A-modules of rank r). More generally, for each open subgroup $K$ of $G(\hat{A})$ , the double coset $G(K)\backslash G(A_f)/K$ may be identified with the set of isomorphism classes of projective A-modules with a level $K$ structure. If $K$ equals $G(\hat{A},n)$ , a level $K$ structure on $Y$ is the choice of an isomorphism $(A/n)^r \xrightarrow{\cong} Y/nY$ .

(1.4) The determinant induces a bijection

$$G(K)\backslash G(A_f)/K \xrightarrow{\cong} K^*\backslash I_f/\det K .$$

This follows for example from the strong approximation theorem for SL(r) [31]. Correspondingly, rank r projective modules over Dedekind domains are, up to isomorphism, uniquely determined by their r-th exterior power [3, VII 4.10].

(1.5) Now we are able to describe the set of C-valued points of the schemes $M_K^r$ of (I 3.3). By (1.3) and (I 2.4), we first get a decomposition by the elements of $P_A^r$ . So let first $Y$ be a lattice in $K^r$ .

$G(K)$ operates from the right on $K_\infty^r$ , and from the left on the set $\mathrm{Mon}(K_\infty^r,C)$ of $K_\infty$-monomorphisms of $K_\infty^r$ into $C$ , by

$$\gamma f(x) = f(x\gamma) \quad (\gamma \in G(K),\ f \in \mathrm{Mon}(K_\infty^r,C)) .$$

We identify $\mathrm{Mon}(K_\infty^r,C)$ with the set

$$\tilde{\Omega}^r = \{\tilde{\omega} = (\omega_1 \dots \omega_r) \in C^r \mid \omega_1 \dots \omega_r \ K_\infty\text{-linearly independent}\} .$$

The linear group $GL(Y)$ of $Y$ is discrete in

$GL(Y \otimes K_\infty) \cong G(K_\infty)$ , and $GL(Y)\backslash Mon(Y \otimes K_\infty, C)$ is canonically isomorphic with the set of those lattices in $C$ which are isomorphic with $Y$ as A-modules. Dividing out the action of $C^*$ , one obtains

$$(1.6) \quad \left\{\begin{matrix}\text{Similarity classes of} \\ \text{lattices isomorphic with } Y\end{matrix}\right\} = GL(Y)\backslash Mon(Y \otimes K_\infty, C)/C^*$$

$$\cong GL(Y)\backslash \tilde{\Omega}^r/C^*$$

$$= GL(Y)\backslash \Omega^r ,$$

where $\Omega^r = \mathbb{P}_{r-1}(C) - \cup \{K_\infty\text{-rational hyperplanes}\}$

in case $r \geq 2$ , and

$\Omega^1$ = point.

This bijection depends on the choice of $Y$ in its class and the identification $Y \otimes K_\infty \cong K_\infty^r$ . In order to get a canonical description, we proceed as follows: On the set $G(A_f) \times \Omega^r$ , $G(K)$ operates from the left by

$$\gamma(\underline{g}, \omega) = (\gamma\underline{g}, \gamma\omega)$$

and $G(A_f)$ from the right by

$$(\underline{g}, \omega)\underline{g}' = (\underline{g}\underline{g}', \omega) ,$$

where $\gamma \in G(K)$, $\underline{g}$, $\underline{g}' \in G(A_f)$ , and $\omega \in \Omega^r$ .

Then we have the <u>canonical</u> bijection

$$(1.7) \quad M_K^r(C) \xrightarrow{\cong} G(K)\backslash G(A_f) \times \Omega^r/K .$$

Choosing a RS $\{\underline{x}\}$ of $G(K)\backslash G(A_f)/K$ , one obtains

$$(1.8) \quad M_K^r(C) \xrightarrow{\cong} \coprod_{\{\underline{x}\}} \Gamma_{\underline{x}}\backslash \Omega^r ,$$

where we have considered $\Gamma_{\underline{x}} = K^{\underline{x}} \cap G(K)$ as a subgroup of $G(K_\infty)$ .

(1.9) Let now $r \geq 2$ . In [11, § 6], Drinfeld defines a structure as a C-analytic manifold on $\Omega^r$ . Discrete subgroups $\Gamma$ of $G(K_\infty)$ operate with finite stabilizers on $\Omega^r$ , and $\Gamma\backslash\Omega^r$ inherits a structure as an eventually singular analytic space. Further, the bijection (1.7) comes from an isomorphism of analytic spaces [11, Prop.6.6]. For us, only the case $r = 2$ will be of interest which will be discussed in detail in (V 1).

## 2. Lattice Invariants

The following computations are taken from [26]. Partially, they trace back to [4,5,6].

Let $\Lambda$ be a r-lattice in $C$ with associated Drinfeld module $\phi$, and let

$$(2.1) \qquad e_\Lambda(z) = \sum \alpha_i z^{q^i}$$

be the lattice function of $\Lambda$. It has a composition inverse

$$(2.2) \qquad \log_\Lambda(z) = \sum \beta_i z^{q^i} .$$

We have

$$(2.3) \qquad \sum_{i+j=k} \alpha_i \beta_j^{q^i} = \sum_{i+j=k} \beta_i \alpha_j^{q^i} = \begin{cases} 1, & \text{if } k = 0 \\ 0 & \text{otherwise.} \end{cases}$$

Let $a$ in $A$ be non-constant, and

$$\phi_a(z) = \sum_{i \le r \cdot \deg a} a_i z^{q^i} .$$

Now

$$e_\Lambda(az) = \phi_a(e_\Lambda(z)) .$$

Applying $\log_\Lambda$ on both sides and substituting $\log_\Lambda$ for $z$ gives

$$(2.4) \qquad a \log_\Lambda(z) = \log_\Lambda(\phi_a(z)) .$$

Equating coefficients, we get for $k \ge 0$

$$(2.5) \qquad a \cdot \beta_k = \sum_{i+j=k} \beta_i a_j^{q^i} .$$

Knowing the $a_i$, one may recursively compute the $\beta_i$, and vice versa. Let further

(2.6) $$z/e_\Lambda(z) = \sum \gamma_i z^i .$$

We are expressing the $\gamma_i$ as lattice sums:

$$\begin{aligned} z/e_\Lambda(z) &= z \sum_{\lambda\in\Lambda} 1/(z-\lambda) && \text{(I 2.2 v)} \\ &= \sum 1/(1-\lambda/z) \\ &= 1 - \sum{}' (z/\lambda)/(1-z/\lambda) \\ &= 1 - \sum_{k\geq 1} E^{(k)}(\Lambda) z^k \end{aligned}$$

with the Eisenstein series of weight $k$

(2.7) $$E^{(k)}(\Lambda) = \sum_{\lambda\in\Lambda}{}' \lambda^{-k} .$$

As a trivial consequence, we get

(2.8) $$\gamma_{i\cdot j} = (\gamma_i)^j ,$$

if $j$ is a power of $p$.

In some cases, we may express the $\gamma_i$ by the $\beta_i$.

2.9 Lemma [26, 2.3.4]: For $j$ of the form $q^k-q^i$, we have

$$\gamma_j = \beta_{k-i}^{q^i} .$$

Proof. By (2.8), we may assume $i = 0$ and argue by induction on $k$, the case $k = 0$ being trivial. For $k > 0$,

$$\sum_{i\leq k} \gamma_{q^k - q^i} \alpha_i = 0$$

by definition of the $\gamma_i$. Therefore,

$$\begin{aligned} \gamma_{q^k - 1} &= - \sum_{1\leq i\leq k} \gamma_{q^k - q^i} \alpha_i \\ &= - \sum \beta_{k-i}^{q^i} \alpha_i && \text{(by induction hypothesis)} \\ &= \beta_k && \text{(by (2.3))} \quad \square \end{aligned}$$

2.10 Conclusion. The lattice $\Lambda$ resp. the Drinfeld module $\phi$ is uniquely determined by the values $E^{(k)}(\Lambda)$. By (2.5), already the knowledge of a finite number of the $E^{(k)}(\Lambda)$ suffices. If we put $E^{(0)}(\Lambda) = -1$, we have the corresponding formula

$$a\,E^{(q^k-1)} = \sum_{i+j=k} E^{(q^i-1)}\,a_j^{q^i}\,. \tag{2.11}$$

## 3. Morphisms of Lattices

The lattice functions of two similar lattices $\Lambda$, $c\Lambda$ in $C$ are related by

$$e_{c\Lambda}(cz) = c\,e_\Lambda(z)\,,$$

and for $a$ in $A$, we have the identity

$$c \cdot \phi_a^\Lambda = \phi_a^{c\Lambda} \circ c \tag{3.1}$$

in $\mathrm{End}_C(G_a)$. (For abbreviation, we write $c \cdot \phi^\Lambda = \phi^{c\Lambda} \circ c$ resp. $\phi^{c\Lambda} = c \cdot \phi^\Lambda \circ c^{-1}$.) Thus, the functions

$$l_i(a,\Lambda) = l_i(\phi_a^\Lambda)$$

are of weight $q^i-1$ on the set of lattices, i.e.

$$l_i(a,c\Lambda) = c^{1-q^i}\,l_i(a,\Lambda)\,. \tag{3.2}$$

For lattices $\Lambda \subset \Lambda'$ in $C$ of the same rank, let $\mu(\Lambda,\Lambda')$ be the corresponding morphism of Drinfeld modules (compare (I 2.3-2.5)). We have

$$e_{\Lambda'} = \mu \circ e_\Lambda \quad \text{and} \quad D(\mu) = 1\,. \tag{3.3}$$

There are several "canonical" normalizations of the polynomial $\mu$ ; for example, we have for $a \in A$

$$a \cdot \mu(\Lambda,a^{-1}\Lambda) = \phi_a^\Lambda\,. \tag{3.4}$$

On the other hand, for an ideal $\mathfrak{n} \subset A$ and $\mu = \mu(\Lambda,\mathfrak{n}^{-1}\Lambda)$,

$$\nu = l^{-1}(\mu) \cdot \mu$$

is the uniquely determined morphism $\phi^{\Lambda} \to \mathfrak{n} * \phi^{\Lambda}$ into the D-module $\mathfrak{n} * \phi^{\Lambda}$ of Hayes [36] which satisfies $l(\nu) = 1$. It corresponds to the lattice multiplication

$$\Lambda \xrightarrow{l^{-1}(\mu)} l^{-1}(\mu)\mathfrak{n}^{-1}\Lambda .$$

If $\underline{n} \in I_f$ generates the ideal $\mathfrak{n}$, the D-module $\mathfrak{n} * \phi^{\Lambda}$ is in the class of $z(\underline{n})(\phi^{\Lambda})$. Here, $z(\underline{n})$ denotes the scalar matrix associated with $\underline{n}$, and we have used the operation of $GL(r, A_f)$ defined in (I 3.2). As is easy to see, for principal ideals (a)

(3.5) $\quad (a) * \phi = l^{-1} \circ \phi \circ l$

holds with $l = l(\phi_a)$.

The following distribution property of the additive polynomials $\phi_a$ is fundamental for the arithmetic of $K$ :

Let $\Lambda \subset C$ be a r-lattice with associated D-module $\phi$, and let $a$ be a non-zero element of $A$. Then we have for each $z \in C$ the identity

(3.6) $\quad l(\phi_a) \prod (X-y) = \phi_a(X-z)$ ,

$y$ on the left side running through the set $\{y \in C \mid \phi_a(y) = \phi_a(z)\}$. Namely, both polynomials in $X$ have the same degree, the same zeroes, the same leading coefficient.

If all the leading coefficients $l(\phi_a)$ lie in some subgroup $S$ of $C^*$, we have for $x \neq 0$

$$\prod_{\phi_a(y)=x} y \equiv x \bmod S .$$

In (IV 4.13), we will use this fact and the logarithm mapping

$$\log_q|\,| : C^* \to \mathbb{Q}$$

to define a $\mathbb{Q}$-valued distribution on the set of division points of a normalized 1-D-module.

For $\mathfrak{a} \subset A$ and

$$\phi_a(X) = l(\phi_a) \prod_{y \in D(\phi,a)} (X-y) \ ,$$

we have the corresponding equation

$$(3.7) \qquad l(\phi_a) \prod_{\phi_a(y)=\phi_a(z)} (X-y) = \phi_a(X-z) \ .$$

## III Partial Zeta Functions

### 1. Relations with Lattice Sums

One of the fundamental facts in the theory of function fields is the famous Theorem of Riemann-Roch:

Let $X$ be a projective, nonsingular, geometrically irreducible algebraic curve of genus $g$ over the perfect field $F$, and $\mathcal{L}$ a line bundle of degree $d$ on $X$. The dimensions $h^i(\mathcal{L})$ of the $F$-vector spaces $H^i(X,\mathcal{L})$ satisfy

$$(1.1) \qquad h^o(\mathcal{L}) - h^1(\mathcal{L}) = 1 - g + d \ .$$

(For a proof in the case $F$ algebraically closed, see [57, Ch.II]. As usual, a line bundle is locally free $\mathcal{O}_X$-sheaf of dimension 1.)

Further, we have the Serre duality

$$(1.2) \qquad H^1(X,\mathcal{L}) \times H^o(X,\mathcal{D} \otimes \mathcal{L}^{-1}) \to F$$

with the canonical line bundle $\mathcal{D}$ of differentials of $X$, i.e. (1.2) is a non-degenerate pairing of finite-dimensional vector spaces. In particular, $h^1(\mathcal{L}) = h^o(\mathcal{D} \otimes \mathcal{L}^{-1})$, and this number vanishes for $d > 2g-2$.

If $\mathcal{L} = \mathcal{L}(D)$ corresponds to the class of the divisor $D$ on $X$, we write $H^i(D)$ for $H^i(X,\mathcal{L}(D))$ etc. We may describe $H^o(D)$ as the $F$-vector space of functions $f$ on $X$ whose divisors satisfy $\operatorname{div}(f) \geq -D$.

Let now $X|\mathbb{F}_q$ be the nonsingular projective curve associated with our function field $K$ and $J$ its Jacobian. We apply (1.1) to divisors of the form $\mathfrak{n} \cdot \infty^i$, $\mathfrak{n}$ denoting a divisor supported by $\operatorname{Spec} A \hookrightarrow X$, i.e. a fractional ideal. Then

$$(1.3)\qquad H^0(\mathfrak{n}^{-1}\infty^i) = \{a \in \mathfrak{n} \mid \deg a \leq i\cdot\delta\} = \mathfrak{n}_{i\delta}\ .$$

The complex zeta function of $K$ is defined by

$$(1.4)\qquad \zeta_K(s) = \sum |\mathfrak{n}|^{-s}\ ,$$

the sum running over the positive divisors of $K$ (eventually possessing an $\infty$-component). The sum converges for real part of $s > 1$ and has a product expansion

$$(1.5)\qquad \zeta_K(s) = \prod_{\mathfrak{p}\text{ place of }K} (1-|\mathfrak{p}|^{-s})^{-1}$$

in this half-plane. Further, $\zeta_K$ is a rational function

$$(1.6)\qquad \zeta_K(s) = Z_K(S) = \frac{P(S)}{(1-S)(1-qS)}$$

in $S = q^{-s}$. The polynomial $P$ satisfies the functional equation $P(X) = q^g X^{2g} P(1/qX)$, we have $P(0) = 1$, and $P(1)$ equals the number $h$ of divisor classes of degree $0$ of $K$ [68, Ch.VII Thm.4].

We are interested in that part of the zeta function which comes from $A$. So let

$$(1.7)\qquad \zeta_A(s) = Z_A(S) = \sum_{\mathfrak{n}\subset A} |\mathfrak{n}|^{-s} = (1-S^\delta)Z_K(S)\ ,$$

and let $\operatorname{Pic} A$ be the group of divisor classes of $A$. We write $\mathfrak{a} \sim \mathfrak{b}$ for the equivalence of fractional ideals, and $(\mathfrak{a})$ for the class of $\mathfrak{a}$ in $\operatorname{Pic} A$. By the principal divisor $(a)$ of $a \in K^*$, we always understand the part prime to the place $\infty$.

For a class $(\mathfrak{a}) \in \operatorname{Pic} A$, let

$$(1.8)\qquad \zeta_{(\mathfrak{a})}(s) = Z_{(\mathfrak{a})}(S) = \sum_{\substack{\mathfrak{n}\subset A\\ \mathfrak{n}\sim\mathfrak{a}}} |\mathfrak{n}|^{-s}\ .$$

$\operatorname{Pic} A$ is described by the short exact sequence

$$(1.9)\qquad 0 \to J(\mathbb{F}_q) \xrightarrow{\text{res}} \operatorname{Pic} A \xrightarrow{\deg} \mathbb{Z}/\delta \to 0\ ,$$

res denoting the restriction to A of a divisor class, and $\deg(\mathfrak{a})$ is the degree $\deg \mathfrak{a} \bmod \delta$ of the fractional ideal $\mathfrak{a}$ .

Instead of summing over divisors, we may use elements of A : For an integral ideal $\mathfrak{n}$ and $a \in A$ , let

$$(1.10) \qquad \zeta_{a,\mathfrak{n}}(s) = Z_{a,\mathfrak{n}}(S) = \sum_{\substack{x \in A \\ x \equiv a \bmod \mathfrak{n}}} |x|^{-s} = \sum S^{\deg x} .$$

Note: These functions are defined for $a \in \mathfrak{n}$ , too. The summand $0^{-s}$ occurring does not contribute.

For these partial zeta functions, the following trivial relations hold:

$$(1.11) \quad \text{(i)} \qquad Z_{a,\mathfrak{n}} = Z_{b,\mathfrak{n}} \text{ , if } a \equiv b \bmod \mathfrak{n} ;$$

$$\text{(ii)} \qquad \sum_{\substack{a \bmod \mathfrak{n}\mathfrak{m} \\ a \equiv b \bmod \mathfrak{n}}} Z_{a,\mathfrak{n}\mathfrak{m}} = Z_{b,\mathfrak{n}} ;$$

$$\text{(iii)} \qquad Z_{ba,b\mathfrak{n}} = S^{\deg b} Z_{a,\mathfrak{n}} \qquad (0 \neq b \in A) ;$$

$$\text{(iv)} \qquad Z_{ca,\mathfrak{n}} = Z_{a,\mathfrak{n}} \qquad (c \in \mathbb{F}_q^*) .$$

Obviously, there is a unique extension of $Z_{*,*}$ to the set of pairs $(a,\mathfrak{n})$ , where $a \in K$ and $\mathfrak{n}$ is a fractional A-ideal, with the same properties. (In (ii), $\mathfrak{m}$ has to be integral, in (iii), we admit $b \in K^*$ .) Properties (i) and (ii) say that $Z_{*,*}$ defines a distribution which is "even" by (iv).

1.12 Lemma. Let $\mathfrak{n}$ be a divisor of degree $d \in \mathbb{Z}$ . Then

$$(q-1) Z_{(\mathfrak{n}^{-1})}(S) = S^{-d} Z_{o,\mathfrak{n}} .$$

Proof.
$$(q-1) Z_{(\mathfrak{n}^{-1})}(S) = (q-1) \sum_{\substack{\mathfrak{a} \subset A \\ \mathfrak{a} \sim \mathfrak{n}^{-1}}} |\mathfrak{a}|^{-s}$$

$$= \sum_{0 \neq f \in \mathfrak{n}} |f\mathfrak{n}^{-1}|^{-s}$$

$$= S^{-d} \sum_{0 \neq f \in \mathfrak{n}} |f|^{-s}$$

$$= S^{-d} Z_{o,\mathfrak{n}} \qquad \square$$

## 2. The Rational Function $Z_{a,\mathfrak{n}}(S)$

We need an explicit an explicit expression of $Z_{a,\mathfrak{n}}(S)$ as a rational function. OBdA we may assume $a$ and $\mathfrak{n}$ to be integral. Let $d = \deg \mathfrak{n}$. We first compute

$$Z_{0,\mathfrak{n}}(S) = \sum_{i \geq 0} \#(\mathfrak{n}_i - \mathfrak{n}_{i-1}) S^i .$$

Since only $i \equiv 0(\delta)$ occur as the degrees of elements of $K$, we define for $i \in \mathbb{Z}$:

(2.1) $\quad i^* = \inf \{n \mid n \geq i,\ n \equiv 0(\delta)\}$

$$i_* = \sup \{n \mid n \geq i,\ n \equiv 0(\delta)\} .$$

In the following, $r$ is always an integer divisible by $\delta$. Obviously,

(2.2) $\quad \dim \mathfrak{n}_r = 1 - g + r - d$, if $r \leq d + 2g - 1$

$\qquad\qquad 0$, $\qquad$ if $r < d$.

In general, $\dim \mathfrak{n}_r = h^o(\infty^{r/\delta} \cdot \mathfrak{n}^{-1})$, and the function $u : \mathbb{Z} \to \mathbb{Z}$ defined by $u(t) = h^o(\infty^{t+d_*/\delta} \cdot \mathfrak{n}^{-1})$ depends only on the class of $\mathfrak{n}$. Put

$$m = m(\mathfrak{n}) = (d+2g-1)^* .$$

Then

$$t \geq (m-d_*)/\delta \iff t\delta + d_* \geq d + 2g - 1$$

and

$$u(t) = 1-g+t\delta+d_*-d, \text{ if } t \geq (m-d_*)/\delta$$

$\qquad\qquad 0$, $\qquad$ if $t < 0$.

A number $t$ $(0 \le t \le (m-d_*)/\delta)$ is called a <u>Weierstraß gap</u> (resp. <u>non-gap</u>) if $u(t) = u(t-1)$ (resp. $u(t) > u(t-1)$) .

(2.3) Let $0 \le t_1 < t_2 \ldots$ the complete sequence of non-gaps, $u_i = u(t_i)$, $v_i = u_i - u_{i-1}$, where we have put $u_0 = 0$ . Evidently, $t_1 = 0$ means $\mathfrak{n}$ is a principal ideal.

2.4 <u>Remark</u>. If $\delta = 1$ , we have $m-d_* = 2g-1$ . Among the $2g$ numbers $t$ with $0 \le t \le 2g-1$ , there are precisely $g$ gaps and non-gaps, each with $v_i = 1$ . There are further restrictions for the set $\{t_i\}$ of non-gaps [49]. Its description differs from that valid in characteristic zero. A priori, if $\delta$ is greater than 1, we have only $v_i \le \delta$ .

For each non-gap $t_i$ , there are precisely $q^{u_i} - q^{u_{i-1}}$ elements $a$ of $\mathfrak{n}$ of degree $\delta t_i + d_*$ . Therefore,

$$(2.5)\qquad Z_{o,\mathfrak{n}}(S) = S^{d_*} \sum_{\{t_i\}} (q^{u_i} - q^{u_{i-1}}) S^{\delta t_i} + \sum_{\substack{r>m \\ r\equiv 0\,(\delta)}} \#(\mathfrak{n}_r - \mathfrak{n}_{r-\delta}) S^r$$

$$= S^{d_*} \sum (q^{u_i} - q^{u_{i-1}}) S^{\delta t_i} + (q^{\delta}-1) q^{1-g-d+m} \frac{S^{m+\delta}}{1 - q^{\delta} S^{\delta}} ,$$

where we have substituted (2.2) and summed up the geometric series. Before computing $Z_{a,\mathfrak{n}}$ , we define for $a \in A$ not in $\mathfrak{n}$

$$(2.6)\qquad r(a) = r(a,\mathfrak{n}) = \inf\{\deg b \mid b \equiv a \bmod \mathfrak{n}\}$$

$$w(a) = w(a,\mathfrak{n}) = \dim \mathfrak{n}_{r(a)} .$$

2.7 <u>Lemma</u>. We have $r(a) \le m$ .

<u>Proof</u>. For $t \in \mathbb{Z}$ , we have the exact sequence

$$\begin{array}{ccccccc} 0 \to & H^o(\infty^t \mathfrak{n}^{-1}) & \to & H^o(\infty^t) & \to A/\mathfrak{n} . \\ & \| & & \| & \\ & \mathfrak{n}_{\delta t} & & A_{\delta t} & \end{array}$$

The right hand map will be surjective if there are no gaps between $\infty^t \mathfrak{n}^{-1}$ and $\infty^t$ . This is surely the case for $\deg(\infty^t \mathfrak{n}^{-1}) \ge 2g-1$ , i.e.

$t \geq m/\delta$ . □

2.8 Lemma. Let $R(r,a) = \{b \in A \mid b \equiv a \bmod \mathfrak{n}$ and $\deg b = r\}$ . Then

$$\#(R(r,a)) = \begin{cases} 0 & r < r(a) \\ q^{w(a)} & r = r(a) \\ q^{\dim \mathfrak{n}_r} - q^{\dim \mathfrak{n}_{r-\delta}} & r > r(a) \,. \end{cases}$$

Proof. Let $a$ be of minimal degree $r(a)$ . For $r \geq r(a)$ , the elements $b$ of $R(r,a)$ are obtained by $b = a+c,\ c \in \mathfrak{n}_r$ . □

Let $Q_i$ be the operator on power series in $S$ which cuts off the terms of degree $\leq i$ . Directly from (2.8), we get

(2.9) $$Z_{a,\mathfrak{n}}(S) = Q_{r(a)} Z_{o,\mathfrak{n}}(S) + q^{w(a)} S^{r(a)} \,.$$

2.10 Corollary. The residues of $Z_{a,\mathfrak{n}}(S)$ do not depend on $a$ .

Proof. By (2.7), $Q_{r(a)}$ does not change the second term of (2.5). □

## 3. Evaluation at $s = 0$ and $s = -1$

Later on, we shall need the values of our zeta fucntions at $s = 0$ and $-1$ , i.e. at $S = 1,q$ . As one sees from (2.5) and (2.9),

(3.1) $$Z_{a,\mathfrak{n}}(1) = \begin{cases} -1 & \text{if } a \in \mathfrak{n} \\ 0 & \text{otherwise.} \end{cases}$$

So, for $S = 1$ , only the derivatives are of interest.

Let now $Z = Z_{a,\mathfrak{n}}$ , and assume for simplicity $a$ and $\mathfrak{n}$ to be integral. We decompose $Z = Z_N + W_N$ , where $W_N = Q_N Z$ , $N \equiv 0(\delta)$ . Then

$$Z_N(S) = \sum_{\substack{x \equiv a \bmod \mathfrak{n} \\ \deg x \leq N}} S^{\deg x} \,, \text{ and}$$

$$(3.2) \qquad Z'_N(1) = \sum_{\substack{x \equiv a \\ \deg x \leq N}} \deg x \ .$$

For sufficiently large $N$ , by (2.5) and (2.9)

$$(3.3) \qquad W_N(S) = (q^\delta - 1)q^{1-g+N-\deg \mathfrak{n}} \frac{S^{N+\delta}}{1-q^\delta S^\delta} \ .$$

Consequently,

$$(3.4) \qquad W'_N(1) = q^{1-g+N-\deg \mathfrak{n}} \left(\frac{\delta}{q^\delta - 1} - N\right) ,$$

and for $d \in \mathbb{N}$ , $d \equiv 0(\delta)$

$$q^d W'_N(1) - W'_{N+d}(1) = dq^{1-g+N+d-\deg \mathfrak{n}} ,$$

which implies the formula

$$(3.5) \qquad q^d Z'_N(1) - Z'_{N+d}(1) + dq^{1-g+N+d-\deg \mathfrak{n}} = (q^d - 1)Z'(1) ,$$

valid for $N >> 0$ .

For any arbitrary Z-function $Z_{*,*}$ and $N \equiv 0(\delta)$ , we correspondingly write $Z_{*,*} = Z_{*,*,N} + W_{*,*,N}$ , where $W_{*,*,N} = Q_N Z_{*,*}$ . By (2.9), the following trivial identity for $N >> 0$ follows:

$$(3.6) \qquad Z'_{a,\mathfrak{n},N}(1) - Z'_{0,\mathfrak{n},N}(1) = Z'_{a,\mathfrak{n}}(1) - Z'_{0,\mathfrak{n}}(1) \ .$$

If $a \notin \mathfrak{n}$ and $b \in K^*$ , (1.12 iii) and (3.1) imply

$$(3.7) \qquad Z'_{ba,b\mathfrak{n}}(1) = Z'_{a,\mathfrak{n}}(1) \ .$$

3.8 Proposition. Let $\mathfrak{n}$ be a fractional ideal. The values $Z'_{a,\mathfrak{n}}(1)$ for $a \in K-\mathfrak{n}$ are bounded above by a constant $Q$ not depending on $a$ and $\mathfrak{n}$ .

Proof. By (3.7), we may assume $a$ and $\mathfrak{n}$ to be integral. Looking at (2.5) and (2.9), one notices: $Z'_{a,\mathfrak{n}}(1)$ takes its maximal value if $w(a,\mathfrak{n})$ is maximal ( $\mathfrak{n}$ supposed to be fixed). But we have $r(a,\mathfrak{n}) \leq m(\mathfrak{n}) = (2g-1+\deg \mathfrak{n})^*$ , and the corresponding maximal value of

$w(\mathfrak{a},\mathfrak{n})$ is $j = 1-g+m-\deg \mathfrak{n} \leq g+\delta-1$ . We obtain

$$Z'_{\mathfrak{a},\mathfrak{n}}(1) \leq \frac{d}{dS}\Big|_{S=1} \left(q^j S^m + (q^\delta -1) q^j \frac{S^{m+\delta}}{1-q^\delta S^\delta}\right)$$

$$= q^j \delta/(q^\delta -1)$$

$$\leq q^{g+\delta-1}\delta/(q^\delta -1) =: Q \qquad \square$$

Now, we examine the values at $S = q$ . We have

(3.9) $$Z_{a,\mathfrak{n},N}(q) = \sum_{\substack{x \equiv a \bmod \mathfrak{n} \\ \deg x \leq N}} q^{\deg x} .$$

Let $\mathfrak{a},\mathfrak{m},\mathfrak{n}$ be integral ideals, $\mathfrak{n} \cdot \mathfrak{m} = (f)$ a principal ideal of degree $d$ . For $N \gg 0$ ,

(3.10) $$q^{-\deg \mathfrak{a}\mathfrak{m}} Z_{o,\mathfrak{a}\mathfrak{m},N+d}(q) - q^{2\deg \mathfrak{n} - \deg \mathfrak{a}} Z_{o,\mathfrak{a},N}(q)$$

$$= q^{-\deg \mathfrak{a}\mathfrak{m}} Z_{o,\mathfrak{a}\mathfrak{m}}(q) - q^{2\deg \mathfrak{n} - \deg \mathfrak{a}} Z_{o,\mathfrak{a}}(q)$$

(the W-parts cancel!)

$$= (q-1)\,[Z_{(\mathfrak{a}^{-1}\mathfrak{m}^{-1})}(q) - q^{2\deg \mathfrak{n}} Z_{(\mathfrak{a}^{-1})}(q)] .$$

Correspondingly, for $a \in \mathfrak{a} - f\mathfrak{a}$ , $u = a/f$ and $N \gg 0$ ,

(3.11) $$q^{-d} Z_{a,f\mathfrak{a},N+d}(q) - Z_{o,\mathfrak{a},N}(q)$$

$$= q^{-d} Z_{a,f\mathfrak{a}}(q) - Z_{o,\mathfrak{a}}(q)$$

$$= Z_{u,\mathfrak{a}}(q) - Z_{o,\mathfrak{a}}(q) ,$$

this number being greater than 0 by (2.9).

## IV Drinfeld Modules of Rank 1

In this chapter, all D-modules are of rank one and, in general, assumed to be defined over $C$ .

Before giving the general description, we first give a sketch of the least complicated case $A = \mathbb{F}_q[T]$ which is strongly analogous with the theory of cyclotomic fields. For a more detailed discussion of this case, see [14,15,16,29,35].

## 1. The Case of a Rational Function Field

Thus let $K = \mathbb{F}_q(T)$ be the field of rational functions, $A = \mathbb{F}_q[T]$ the ring of polynomials, and "$\infty$" the place at infinity of $K$. We have the distinguished uniformizing parameter $\pi = T^{-1}$ at $\infty$, and we call $x \in K_\infty$ <u>monic</u> if it is non-zero and has the leading coefficient 1 in its $\pi$-expansion.

A D-module $\phi$ is given by

$$\phi_T = T\tau^0 + l\tau$$

with $l \neq 0$, and is isomorphic with the Carlitz module

(1.1) $$\rho_T = T\tau^0 + \tau ,$$

corresponding to the fact that all the 1-lattices in $C$ are similar. The zeroes $x$ of $\rho_a$ ($a \in A$ non-constant, $\rho_a$ now considered as an additive polynomial) generate an abelian extension $K(a)$ of $K$. We have

(1.2) $$G(A/a) = (A/a)^* \xrightarrow{\cong} \mathrm{Gal}(K(a) : K)$$

$$b \longmapsto \sigma_b ,$$

where $\sigma_b$ operates by $\sigma_b(x) = \rho_b(x)$ on the zero $x$ of $\rho_a$. Let $K_+(a)$ be the fixed field of $\mathbb{F}_q^* \subset G(A/a)$.

1.3 <u>Theorem</u> [35]. (i) The place $\infty$ is totally split in $K_+(a)$ and totally ramified in $K(a) : K_+(a)$.

(ii) If $a = \prod p_i^{e_i}$ is the prime decomposition, $K(a)$ is the compositum of the linearly disjoint fields $K(p_i^{e_i})$.

(iii) $K(p_i^{e_i})$ is ramified at most at $\infty$ and $(p_i)$, the ramification at $(p_i)$ being total.

Decomposing the idele group of $K$

$$I = I_f \times K_\infty^* ,$$

$$K_\infty^* = E_\infty^{(1)} \times \mu_{q-1} \times \pi^{\mathbb{Z}}$$

with the group $E_\infty^{(1)}$ of 1-units, $K(a)$ corresponds to the norm group $(E_f(a) \times E_\infty^{(1)} \times \pi^{\mathbb{Z}}) \cdot K^*$, and $K_+(a)$ to the norm group $(E_f(a) \times K_\infty^*) \cdot K^*$ [1, Ch.8].

For the first class field theory of $K$, the Carlitz module plays the same role as does the multiplicative group scheme for the field $\mathbb{Q}$. In fact, the analogy is much deeper than appears at first sight. (See [14,15] for a relation between unit indices and class numbers of subfields of $K(a)$, for example.) By obvious reasons, the $K(a)$ are called "cyclotomic extensions" of $K$.

The Carlitz module $\rho$ is associated with a lattice $\Lambda = \xi \cdot A$, the period $\xi$ being defined up to a $(q-1)$-st root of unity.

1.4 Theorem [4,26]. We have

$$\xi^{q-1} = (T-T^q)\xi_0^{q-1}, \text{ where}$$

$$\xi_0 = \prod_{i\geq 1} \frac{1-\pi^{(q-1)q^i}}{1-\pi^{(q^{i+1}-1)}} .$$

1.5 Remark. In the above mentioned analogy of $K$ with $\mathbb{Q}$, $\xi_0$ corresponds to the number $\pi$ and $\xi$ to $2\pi i$. Several further expressions for $\xi$ are known, for instance

$$\xi^{q-1} = (T^q-T) \sum_{a\in A}{}' a^{1-q} ,$$

or the product (4.10) over $A$.

## 2. Normalization [39, § 4]

Let now $(K,\infty)$ again be arbitrary. The residue field $k$ at $\infty$ is of degree $\delta$ above $\mathbb{F}_q$, and has a uniquely determined lifting $k \hookrightarrow K_\infty$. Recall that $E_\infty$ resp. $E_\infty^{(1)}$ is the group of units resp. of 1-units of $K_\infty$.

2.1 Definition. A sign-function is a map $\mathrm{sgn} : K_\infty \to \mathbb{k}$ with the properties

(i) $\mathrm{sgn}(xy) = \mathrm{sgn}(x)\,\mathrm{sgn}(y)$ ;

(ii) $\mathrm{sgn}(x) = 1 \quad (x \in E_\infty^{(1)})$ ;

(iii) $\mathrm{sgn}(x) = x \quad (x \in \mathbb{k})$ .

A sign function is uniquely determined by its values on $A$ . There exist precisely $w = q^\delta - 1$ sign functions. They correspond to non-zero tangent vectors at the point $\infty$ . A twisted sign function is a function of the form $\tau^i \circ \mathrm{sgn}$ with a sign function sgn and some $\tau^i \in \mathrm{Gal}(\mathbb{k} : \mathbb{F}_q)$ . We choose a fixed sign function sgn and a parameter $\pi$ at $\infty$ with $\mathrm{sgn}(\pi) = 1$ , and we call $x \in K_\infty$ monic if $\mathrm{sgn}(x) = 1$ . We shall see that the choice of $\pi$ is of no importance.

We would like to have, for each similarity class of 1-lattices in $C$ , i.e. for each ideal class of $A$ , a "canonical" Drinfeld module whose coefficients are as simple as possible. First, one proves straightforward:

2.2 Lemma. If all the leading coefficients of the D-module $\phi$ lie in $\mathbb{k}$ , the map $a \longmapsto 1(\phi_a)$ is a twisted sign function.

2.3 Definition. The D-module $\phi$ is called normalized (resp. sgn-normalized), if the map $a \longmapsto 1(\phi_a)$ takes values in $\mathbb{k}$ (resp. agrees with sgn up to Galois twist).

Further, the following observation is useful:

2.4 Lemma [36, 7.4] . Let $\phi$ be a 1-D-module over a discretely valued field $L$ . If all the leading coefficients of $\phi$ are units, $\phi$ has coefficients in the ring of integers of $L$ .

(2.5) Again, we decompose the idele group

$$I = I_f \times K_\infty^*$$

$$K_\infty^* = E_\infty^{(1)} \times \mu_w \times \pi^{\mathbb{Z}} ,$$

and we define, for each proper A-ideal $\mathfrak{n}$ , the following abelian field extensions of $K$ (considered as subfields of $C$ ) by their norms in the idele class group of $K$ (see [1, Ch.8]):

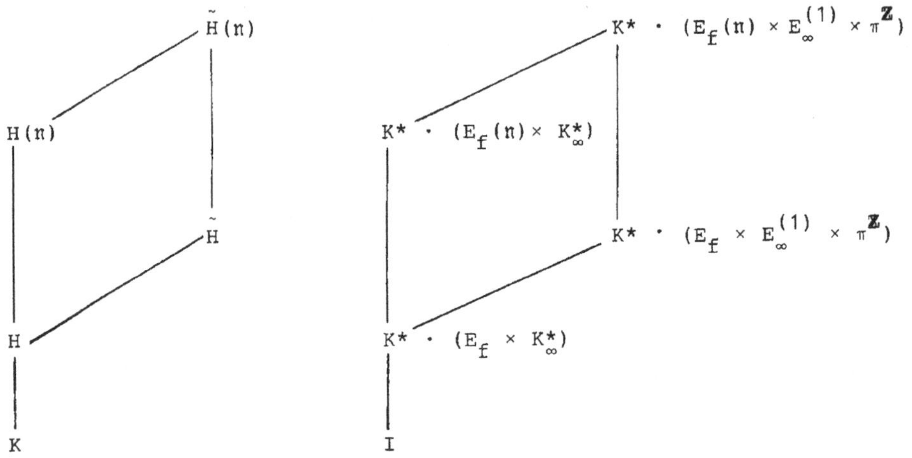


Let further $B$, $\tilde{B}$, $B(\mathfrak{n})$, $\tilde{B}(\mathfrak{n})$ the rings of integers (i.e. the integral closures of $A$ ) in $H$, $\tilde{H}$, $H(\mathfrak{n})$, $\tilde{H}(\mathfrak{n})$ .

Then

(2.6) (i) $H : K$ is unramified and splits completely at $\infty$ . We have $\mathrm{Gal}(H : K) \xrightarrow{\cong} I_f/E_f \cdot K^* \xrightarrow{\cong} \mathrm{Pic}\, A$ . $H$ is called the <u>Hilbert class field</u> of $(K,\infty)$ resp. $A$ .

(ii) $\tilde{H} : H$ is unramified at finite places and totally ramified above $\infty$ . We have $\mathrm{Gal}(\tilde{H} : H) \xrightarrow{\cong} \Bbbk^*/\mathbb{F}_q^*$ . $\tilde{H}$ is the <u>normalizing field</u> of $(K,\infty,\mathrm{sgn})$ resp. $(A,\mathrm{sgn})$ .

(iii) $\tilde{H}(\mathfrak{n}) : \tilde{H}$ is unramified outside of $\mathfrak{n}$ and $\infty$ , and $\mathrm{Gal}(\tilde{H}(\mathfrak{n}) : \tilde{H}) \xrightarrow{\cong} (A/\mathfrak{n})^*$ .

(iv) $\mathrm{Gal}(\tilde{H}(\mathfrak{n}) : H(\mathfrak{n})) \xrightarrow{\cong} \Bbbk^*$ is the inertia group = decomposition group of $\infty$ in $\mathrm{Gal}(\tilde{H}(\mathfrak{n}) : K)$ .

Let $\widetilde{\mathrm{Pic}}\, A$ be the <u>narrow ideal class group</u> of $A$ with respect to sgn. ( $\mathfrak{a}$ and $\mathfrak{b}$ define the same class in $\widetilde{\mathrm{Pic}}\, A$ , if $\mathfrak{a} = f\mathfrak{b}$ with some monic $f \in K^*$ .) Then $\mathrm{Gal}(\tilde{H} : K) \xrightarrow{\cong} \widetilde{\mathrm{Pic}}\, A$ . Further, $\Bbbk$ is the constant field of $H \ldots \tilde{H}(\mathfrak{n})$ . As a first step, we have

2.7 <u>Theorem</u> [36, §§ 6,8]. Each 1-D-module $\phi$ over $C$ is isomorphic with some $\phi'$ with coefficients in $H$ , and $H$ is minimal with this property.

Now, (I 4.1) suggests to look for such a $\phi'$ over $B$ . However, for

$\delta > 1$ , such a normalization does not exist, see below.

In the following, $\phi$ is a 1-D-module over $H$ .

2.8 <u>Proposition</u> [36, 10.3]. There exists $s \in C$ with $s^w \in H$ such that $\phi' = s \circ \phi \circ s^{-1}$ is sgn-normalized.

The number $s$ is uniquely determined up to w-th roots of unity. The corresponding $\phi'$ has coefficients in $H(s^{q-1})$ , and this is the smallest field over which a sgn-normalized $\phi'$ isomorphic with $\phi$ may be defined. Now (2.4) implies

2.9 <u>Corollary</u>. If $\delta$ equals 1, $\phi'$ has coefficients in $B$ . Each 1-D-module over $C$ is isomorphic with such a $\phi'$ .

The general case will be described by

2.10 <u>Theorem</u> [39, § 4]. The field $H(s^{q-1})$ equals the normalizing field $\tilde{H}$ of $(A, \mathrm{sgn})$ . In particular, it does not depend on $\phi$ .

The field $k$ being the constant field of $\tilde{H}$ , (2.4) gives

2.11 <u>Corollary</u>. For each $\phi$ over $C$ , there exists a sgn-normalized $\phi'$ over $\tilde{H}$ isomorphic with $\phi$ . It has coefficients in $\tilde{B}$ , and $\tilde{H}$ is the smallest field with this property.

Let now $\phi$ be sgn-normalized over $\tilde{H}$ , and let $\mathfrak{n} \subsetneq A$ . The field extension $\tilde{H}(D(\phi,\mathfrak{n})) : \tilde{H}$ is abelian with a subgroup of $\mathrm{Aut}_A(D(\phi,\mathfrak{n})) = (A/\mathfrak{n})^*$ as its Galois group. Decomposing $\mathfrak{n}$ into primary components and looking at the ramification properties (see(1.3)), it is not hard to show:

2.12 <u>Theorem</u>. $\tilde{H}(D(\phi,\mathfrak{n}))$ is the field $\tilde{H}(\mathfrak{n})$ , independently of $\phi$ . The group

$$(A/\mathfrak{n})^* \xrightarrow{\cong} \mathrm{Gal}(\tilde{H}(\mathfrak{n}) : \tilde{H})$$

$$a \longmapsto \sigma_a$$

acts on $D(\phi,\mathfrak{n})$ by $\sigma_a(x) = \phi_a(x)$ .

The 1-lattice $\Lambda$ in $C$ is called <u>special</u> if its associated D-module $\phi^\Lambda$ is sgn-normalized. By (2.11), we get at once the

2.13 <u>Corollary</u>. Each similarity class of 1-lattices contains special lattices. These are all conjugate by $k^* = \mu_w$ .

Provisionally, we define the <u>invariant</u> $\xi(\Lambda)$ <u>of $\Lambda$</u> by the condition

(2.14) $\xi(\Lambda)\cdot\Lambda$ is special.

Thereby, $\xi(\Lambda)$ is determined up to multiplication by elements of $\mu_w$. Later on (5.1), we shall choose a specific value for $\xi$.

## 3. Some Lemmata

In this section, some definitions and lemmata needed for the computation of $\xi(\mathfrak{a})$ are collected.

Each $x \in K_\infty^*$ may be decomposed uniquely

(3.1) $x = \mathrm{sgn}(x)\langle x\rangle\pi^i$

with $\langle x\rangle$ a 1-unit and $i \in \mathbb{Z}$.

Let $\mathfrak{a} \subset A$ be an ideal and $f \in A$ of degree $d > 0$. We choose a section of the $\mathbb{F}_q$-linear map $\mathfrak{a} \to \mathfrak{a}/f\mathfrak{a}$, and we use the image set $\{c\}$ as a RS for $\mathfrak{a}/f\mathfrak{a}$. We may choose the elements $c$ of degree $\leq m$, where $m = (2g-1+d+\deg\mathfrak{a})^*$ (compare (III 2.7)).

$N$ will always denote a natural number divisible by $\delta$, and all the following limits refer to $N \to \infty$. We observe that for $N >> 0$, the mapping

(3.2)
$$\mathfrak{a}_N \times \mathfrak{a}/f\mathfrak{a} \longrightarrow \mathfrak{a}_{N+d}$$
$$(a,c) \longmapsto af + c = b$$

is bijective. Each $a$ of a precise degree $N$ corresponds to $q^d$ elements $b$ of degree $N+d$, and $\mathrm{sgn}(b) = \mathrm{sgn}(af+c) = \mathrm{sgn}(a)\mathrm{sgn}(f)$.

3.3 <u>Lemma</u>. For large $N$, the product $\varepsilon_N = \prod \mathrm{sgn}(b)$, where $b$ runs over the elements of $\mathfrak{a}$ of degree $N$, does not depend on $N$. The limit $\varepsilon_{(\mathfrak{a})}$ is an invariant of the ideal class $(\mathfrak{a})$ of $\mathfrak{a}$.

<u>Proof</u>. From (3.2) and $\delta|d$, we get the d-periodicity of $\varepsilon_N$ for $N >> 0$. Now, the first assertion follows from $\gcd\{d|d=\deg f,\ f\in A\} = \delta$. The second assertion comes out by collecting constant factors in the products $\varepsilon_N$, the number $\#(\mathfrak{a}_N - \mathfrak{a}_{N-\delta})$ of these factors being divisible by $w$. □

If $\delta$ equals 1, $\varepsilon_{(a)}$ will always take the value $-1$. But already in the case $K = \mathbb{F}_q(T)$, $\delta > 1$, $\varepsilon_{(*)}$ is rather complicated to describe.

Let $u \in K - \mathfrak{a}$ with $fu \in \mathfrak{a}$. By (3.3), the limit

$$(3.4) \qquad \varepsilon_{u,\mathfrak{a}} = \lim \prod_{\substack{b \in K \\ b \equiv u \bmod \mathfrak{a} \\ \deg b \leq N}} \operatorname{sgn}(b) / \prod_{a \in \mathfrak{a}_N}{}' \operatorname{sgn}(a)$$

exists, and for non-zero $t \in A$,

$$\varepsilon_{tu,t\mathfrak{a}} = \operatorname{sgn}(t)\,\varepsilon_{u,\mathfrak{a}},$$

since the numerator of $\varepsilon_{u,\mathfrak{a}}$ contains one more factor than the denominator.

Next, we examine the behavior of the 1-units. Let

$$U_N = \prod_{\substack{a \in \mathfrak{a} \\ \deg a = N}} \langle a \rangle .$$

3.5 Lemma. $U_N$ converges to 1.

Proof. Consider the product $\prod \langle b \rangle = \prod_{\{c\}} \langle af+c \rangle$ for a fixed $a$ of degree $N \gg 0$ (notations as in (3.2)). We have $\langle af+c \rangle = \operatorname{sgn}(af)^{-1} \pi^{(N+d)/\delta}(af+c)$, so

$$\prod \langle b \rangle = \operatorname{sgn}(af)^{-1} \pi^{q^d (N+d)/\delta} \cdot H(af),$$

$H$ denoting the additive polynomial

$$H(X) = \prod_{c} (X-c) = X^{q^d} + h_{d-1} X^{q^{d-1}} + \ldots h_0 X .$$

Putting $m = (2g-1+d+\deg \mathfrak{a})^*$, we get the trivial upper bound $m(q^d-1)$ for the degree of the coefficients $h_i$. (As already mentioned, $m$ is an upper bound for all the $\deg c$.) Thus

$$(*) \qquad \prod \langle b \rangle = \operatorname{sgn}(af)^{-1} \pi^{q^d (N+d)/\delta} (af)^{q^d} + \text{terms of order}$$
$$\geq (N+d)(q^d - q^{d-1})/\delta - m(q^d-1)/\delta \quad \text{in } \pi .$$

Let now $n \in \mathbb{N}$ be given. We show: There is $M_o \in \mathbb{N}$ such that $U_M \equiv 1 \bmod \pi^n$ holds for all $M \geq M_o$. In proving the lemma, $f$ is at our disposal. So choose $f$ of degree $d$ with $q^d \geq n$. Choose the RS $\{c\}$ for $\mathfrak{a}/f\mathfrak{a}$ such that all the deg $c \leq m = (2g-1+d+\deg \mathfrak{a})^*$. Choose $N$ large enough such that $[(N+d)(q^d-q^{d-1}) - m(q^d-1)]/\delta \geq n$, and put $M_o = N+d$. The product $\prod \langle b \rangle$ over all the $b \in \mathfrak{a}$ of degree $N+d$ decomposes into partial products $(*)$, and it will suffice to show all of them to be $\equiv 1 \bmod \pi^n$.

The first term in $(*)$ is the 1-unit $[\mathrm{sgn}(af)^{-1} af\, \pi^{(N+d)/\delta}]^{q^d}$ which is $\equiv 1 \bmod \pi^n$ by $q^d \geq n$, and the other terms vanish $\bmod\ \pi^n$. Obviously, $U_M \equiv 1 \bmod \pi^n$ is also true for $M \geq M_o$. □

By the lemma just proved, the limit

(3.6) $$U_{o,\mathfrak{a}} = \lim \prod_{a \in \mathfrak{a}_N}{}' \langle a \rangle$$

exists, as well as

$$U_{u,\mathfrak{a}} = \lim \prod_{\substack{b \equiv u \bmod \mathfrak{a} \\ \deg b \leq N}} \langle b \rangle \qquad (u \in K-\mathfrak{a}) .$$

These limits depend on the choice of the parameter $\pi$ at $\infty$. However, the property $\lim u^{q^N} = 1$ of 1-units $u$ allows to control what happens if one changes $\pi$, see e.g. (4.10), (4.12).

3.7 <u>Lemma</u>. The numbers $U_{u,\mathfrak{a}} \in E_\infty^{(1)}$ satisfy

(i) $U_{u,\mathfrak{a}} = U_{v,\mathfrak{a}}$, if $u \equiv v \bmod \mathfrak{a}$;

(ii) $$\prod_{\substack{u \bmod \mathfrak{a}\mathfrak{b} \\ u \equiv v \bmod \mathfrak{a}}} = U_{v,\mathfrak{a}} ;$$

(iii) if $0 \neq t \in A$, $$U_{tu,t\mathfrak{a}} = \begin{cases} U_{u,\mathfrak{a}} & u \notin \mathfrak{a} \\ \langle t \rangle^{-1} U_{u,\mathfrak{a}} & u \in \mathfrak{a} ; \end{cases}$$

(iv) $U_{cu,\mathfrak{a}} = U_{u,\mathfrak{a}}$ for $c \in \mathbb{F}_q^*$.

<u>Proof</u>. Everything except (iii) is obvious. But $U_{tu,t\mathfrak{a}} = \lim \langle t \rangle^{i(N)} \cdot U_{u,\mathfrak{a}}$, where $i(N) = \#(\mathfrak{a}_N)$ in case $u \notin \mathfrak{a}$ and $i(N) = \#(\mathfrak{a}_N) - 1$ otherwise,

and $<t>^{\#(a_N)}$ converges to 1. □

3.8 Remarks. (i) Obviously, the integrality condition on $a$ is unnecessary, and was used only to simplify the notation. By means of (3.7), one ontains an extension of $U_{*,*}$ to the set of all pairs $(u,a)$ with the same properties, compare (III 1.11).

(ii) The properties of $U_{*,*}$ described in (3.7) coincide with the properties of the derivatives $Z'_{*,*}(1)$ of the Z-functions at $S = 1$. Perhaps, the $U_{*,*}$ arise as some sort of derivative of the zeta functions defined by David Goss [30]. But, up to now, there is no differential calculus of these functions. (Just recently, there has been discovered a relation with the $\Gamma$-function of $(K,\infty)$, see forthcoming work of David Goss and Dinesh Thakur.)

## 4. Computation of Lattice Invariants

Now, we compute $\xi(a)$ for ideals $a$ of $A$ considered as 1-lattices in $C$, and use the result to fix $\xi(a)$ (which a priori is defined only up to w-th roots of unity). We keep the notations of the last section. Let $r$ be the degree of $a$, $\Lambda$ the lattice $\omega a$ in $C$ with $\omega$ variable, and $\phi = \phi^{\Lambda}$ the corresponding D-module. Let further $\{u\}$ the RS $\{c/f\}$ of $f^{-1}a/a$. First, some simple number theoretic observations:

(4.1) If $S$ is a subset of $\mathbb{N}$ with $t = \gcd(S)$, then $q^t-1 = \gcd\{q^s-1 \mid s \in S\}$.

(Consider the associated finite fields!)

4.2 Lemma. Let $x_i \in C^*$ be finitely many numbers with weights $g_i$ prime to the characteristic $p$ of $C$, having the property

$$\prod x_i^{r_i} = 1 \quad \text{whenever} \quad \sum r_i g_i = 0 .$$

Then there exists a well defined number $y \in C$ of weight $k = \gcd\{g_i\}$ such that $x_i = y^{g_i/k}$, and $y$ lies in the multiplicative group generated by the $x_i$.

Proof (by induction on the number $n$ of the $x_i$):

n = 2 : We have two elements $x$, $x'$ of weights $g$, $g'$. OBdA, let

$k = (g,g') = 1$ (otherwise, replace $g$, $g'$ by $g/k$, $g'/k$ ). By assumption, $x^{g'} = x'^{g}$ . First, put $\tilde{y} = (g\cdot g')$-th root of $x^{g'}$ . Using the isomorphism

$$\mu_{gg'} \xrightarrow{(g,g')} \mu_{g'} \times \mu_{g} \ ,$$

we get a $(g\cdot g')$-th root of unity $\varepsilon$ such that $y = \varepsilon\cdot\tilde{y}$ satisfies $y^{g} = g$ , $y^{g'} = x'$ . If $sg + tg' = 1$ , then $x^{s}x'^{t} = y$ .

<u>$n > 2$</u> : Let $k' = \gcd \{g_1,\ldots,g_{n-1}\}$ . By induction hypothesis, there exists an $y'$ of weight $k'$ satisfying the assertion of the lemma. Now, apply the case $n = 2$ to $\{y',x_n\}$ ! □

We return to the computation of $\xi(\mathfrak{a})$ . The functions

$$e_u(\omega) = e_\Lambda(u\omega) \ ,$$

where $u$ runs through our RS of $f^{-1}\mathfrak{a}/\mathfrak{a}$ , satisfy

$$e_u(\omega) = \omega \ e_u(1) \ . \tag{4.3}$$

For $\omega$ fixed, they are the zeroes of

$$\phi_f = \sum_{0\leq i\leq d} l_i\tau^i \ ,$$

where $d = \deg f$ . Using $l_o = f$ , we get for $l = l_d$

$$l = f \prod_u{}' \ e_u^{-1} \ . \tag{4.4}$$

We compute

$$\begin{aligned}\prod_u{}' \ e_u(\omega) &= \prod_u{}'(u\omega \prod_{a\in\mathfrak{a}}{}'(1-\frac{u\omega}{a\omega})) \\ &= f^{1-q^d}\omega^{q^d-1}(\prod_c{}'c) \cdot \prod_c{}'(\prod_a{}'\frac{af-c}{af}) \ ,\end{aligned} \tag{4.5}$$

now taking the product over the RS $\{c\}$ of $\mathfrak{a}/f\mathfrak{a}$ . We may interchange the order of product and take the partial product over the $a \in \mathfrak{a}_N$ . Let $N >> 0$ . Then

$$(4.6) \qquad f^{-q^d} (\prod_c{}' c) \prod_c{}' ( \prod_{a \in \mathfrak{a}_N}{}' \frac{af-c}{af} )$$

$$= f^{-q^d} \prod_{b \in \mathfrak{a}_{N+d}}{}' b \Big/ \prod_{a \in \mathfrak{a}_N}{}' (af)^{q^d} ,$$

for the $af-c$ with $c \neq 0$ run through $\mathfrak{a}_{N+d} - \{c\}$ , and in the second product over $\{c\}$ , we may admit $c = 0$ without changing anything.

Substituting (3.1) for a,b,f into (4.6) and collecting similar terms, this becomes

$$(4.7) \qquad \mathrm{sgn}(f)^{-j} \cdot \varepsilon \cdot \langle f\rangle^{-j} \pi^k \prod \langle b\rangle / \langle a\rangle^{q^d} , \text{ where}$$

$$j = q^d(\#(\mathfrak{a}_N)-1) + q^d = q^{1-g-r+N+d} ,$$

$$\varepsilon = \prod{}' \mathrm{sgn}(b) / \prod{}' \mathrm{sgn}(a) ,$$

$$k = [q^d \textstyle\sum' \deg a - \sum' \deg b + dj]/\delta ,$$

b running through $\mathfrak{a}_{N+d}$ and a through $\mathfrak{a}_N$ . We have further used $\mathrm{sgn}(x)^{q^d} = \mathrm{sgn}(x)$ .

In the limit $N \to \infty$ , we obtain

$$\lim \langle f\rangle^j = 1 , \ \langle f\rangle \text{ being a 1-unit,}$$

$$\lim \mathrm{sgn}(f)^j = \tau^{1-g-r}(\mathrm{sgn}(f)) ,$$

$$\lim \varepsilon = \varepsilon_{(\mathfrak{a})}^{d/\delta} \qquad (3.3) ,$$

$$\lim k = (q^d-1) Z'_{o,\mathfrak{a}}(1)/\delta \qquad (\text{III } 3.5) ,$$

$$\lim \prod{}' \langle b\rangle / \prod{}' \langle a\rangle^{q^d} = U_{o,\mathfrak{a}}^{1-q^d} \qquad (3.6) .$$

Substituting the limits into (4.7) and combining (4.4)-(4.7) gives

$$(4.8) \qquad l(\omega) = \tau^{1-g-r}(\mathrm{sgn}(f)) \omega^{1-q^d} \varepsilon_{(\mathfrak{a})}^{-d/\delta} \pi^{-k} U_{o,\mathfrak{a}}^{q^d-1} ,$$

$$k = (q^d-1) Z'_{o,\mathfrak{a}}(1)/\delta .$$

$\phi$ is sgn-normalized if for each $f$ the leading coefficient $l(\phi_f)$ of $\phi_f$ lies in $k$, and the function $f \longmapsto l(\phi_f)$ is of the form $\tau^i \circ \mathrm{sgn}$. This gives the condition for the invariant $\xi = \xi(\mathfrak{a})$ :

$$l(\phi_f, \xi) = 1 \ , \text{ if } \ \mathrm{sgn}(f) = 1 \ .$$

By (4.8),

$$(4.9) \qquad \xi^{q^d-1} = \varepsilon_{(\mathfrak{a})}^{-d/\delta} \ \pi^{-k} \ U_{o,\mathfrak{a}}^{q^d-1} \ .$$

But gcd $\{d|d = \deg f, f \text{ monic}\} = \delta$. Thus, by (4.1) and (4.2), (4.9) already determines $\xi^w$, where $w = q^\delta-1$. Taking into account $\varepsilon^{(q^d-1)/w} = \varepsilon^{d/\delta}$ for $w$-th roots of unity $\varepsilon$, we finally obtain

$$(4.10) \qquad \xi^w = \varepsilon_{(\mathfrak{a})}^{-1} \ \pi^{-k} \ U_{o,\mathfrak{a}}^w \ , \text{ where}$$

$$k = w \cdot Z'_{o,\mathfrak{a}}(1)/\delta \ .$$

In particular, the absolute value is given by

$$|\xi| = q^{Z'_{o,\mathfrak{a}}(1)} \ .$$

Let us now substitute a not necessarily monic $f$ into (4.8). From (4.10),

$$l(\phi_f, \xi) = \tau^{1-g-r}(\mathrm{sgn}(f)) \ ,$$

i.e. the twisted sign function $f \longmapsto l(\phi_f)$ is

$$(4.11) \qquad f \longmapsto \tau^i(\mathrm{sgn}(f)) \ ,$$

where $i = 1-g-r = 1-g-\deg \mathfrak{a}$. Therefore, we put $\mathrm{sgn}(f,\mathfrak{a}) = \tau^{1-g-\deg \mathfrak{a}}(\mathrm{sgn}(f))$.

The corresponding computations for $e_u(\omega)$ $(u \neq 0)$ lead to

$$e_u(\xi) = \xi \ e_u(1)$$

$$= -\xi \lim \pi^{-k} \prod_b \mathrm{sgn}(b) / \prod_a{}' \mathrm{sgn}(a) \cdot \prod_b \langle b \rangle \prod_a{}' \langle a \rangle \ ,$$

$$\delta k = \sum \deg b - \sum{}' \deg a \ ,$$

$b$ running through $\{b \in K \mid \deg b \leq N,\ b \equiv -u \bmod \mathfrak{a}\}$ and $\mathfrak{a}$ through $\mathfrak{a}_N$. Combined with (III 3.6) and the computations of the last section, we get

$$(4.12)\qquad e_u(\xi) = -\xi\, \varepsilon_{-u,\mathfrak{a}}\, \pi^{-k}\, U_{-u,\mathfrak{a}}\, U_{o,\mathfrak{a}}^{-1}$$

$$= \xi\, \varepsilon_{u,\mathfrak{a}}\, \pi^{-k}\, U_{u,\mathfrak{a}}\, U_{o,\mathfrak{a}}^{-1},$$

$$k = [Z'_{u,\mathfrak{a}}(1) - Z'_{o,\mathfrak{a}}(1)]/\delta .$$

4.13 <u>Corollary</u>. The absolute value of $e_u(\xi)$ is given by

$$|e_u(\xi)| = q^{Z'_{u,\mathfrak{a}}(1)} .$$

In particular, $|e_u(\xi)|$ is bounded above by $q^Q$ with the constant $Q$ of (III 3.8).

4.14 <u>Remarks</u>. (i) By a general theorem of Yu on the values of the lattice functions $e_\Lambda$ [71,72], the $\xi(\mathfrak{a})$ are transcendental over $K$.

(ii) (4.13) is Theorem 6.1 of Hayes in [39]. It is the essential step in proving the function field analogue of Stark's abelian conjectures [66, Ch.V]. Compared to the proof originally given by Deligne and Tate, the advantage of this approach is the explicit construction of functions with prescribed divisors.

(iii) (4.10) is analogous with the product expansion

$$\pi = 2 \prod_{a \geq 1} (1-1/4a^2)^{-1}$$

$$= 2 \lim_{N\to\infty} \prod_{\substack{a\in\mathbb{Z}\\ |a|\leq N}}{}' (1-1/2a)^{-1}$$

derived from the product expansion of $\sin \pi z$.

(iv) By means of (II 2), we may derive additive expansions for $\xi^{q^d-1}(\mathfrak{a})$, for example (1.5). In general, $\xi^{q^d-1}$ is a polynomial of weight $q^d-1$ in the lattice sums

$$E^{(q^i-1)}(\mathfrak{a}) = \sum_{a\in\mathfrak{a}}{}' a^{1-q^i} .$$

## 5. Distinguished 1-D-Modules

Up to now, the invariant $\xi(\Lambda)$ of a 1-lattice $\Lambda$ in $C$ was determined only up to w-th roots of unity. Now we choose, for each ideal class $(\mathfrak{a})$ of $A$, a special lattice $\Lambda^{(\mathfrak{a})}$ with associated D-module $\rho^{(\mathfrak{a})}$, and we define the invariant of $\Lambda$ by

$$(5.1) \qquad \xi(\Lambda) \cdot \Lambda = \Lambda^{(\mathfrak{a})} ,$$

provided that $\Lambda$ is similar to $\mathfrak{a}$. This determines $\xi$ at least up to (q-1)-st roots of unity, which will suffice for our purposes. For an integer ideal $\mathfrak{a}$ of $A$, let

$$\rho_{\mathfrak{a}}^{(\mathfrak{b})} : \rho^{(\mathfrak{b})} \to \mathfrak{a} * \rho^{(\mathfrak{b})}$$

be the uniquely determined morphism of (II 3.4) with $1(\rho_{\mathfrak{a}}^{(\mathfrak{b})}) = 1$. This corresponds to the lattice morphism

$$\Lambda^{(\mathfrak{b})} \hookrightarrow \mathfrak{a}^{-1}\Lambda^{(\mathfrak{b})} \xrightarrow{1^{-1}(\mu)} 1^{-1}(\mu)\mathfrak{a}^{-1}\Lambda^{(\mathfrak{b})}$$

with $\mu = \mu(\Lambda^{(\mathfrak{b})}, \mathfrak{a}^{-1}\Lambda^{(\mathfrak{b})})$. We have $D(\rho_{\mathfrak{a}}^{(\mathfrak{b})}) = 1^{-1}(\mu)$. The D-module $\mathfrak{a} * \rho^{(\mathfrak{b})}$ is isomorphic with $\rho^{(\mathfrak{a}^{-1}\mathfrak{b})}$. Comparing the leading coefficients shows $\mathfrak{a} * \rho^{(\mathfrak{b})}$ to be sgn-normalized. Thus, an isomorphism

$$(5.2) \qquad \theta(\mathfrak{a},\mathfrak{b}) : \mathfrak{a} * \rho^{(\mathfrak{b})} \xrightarrow{\cong} \rho^{(\mathfrak{a}^{-1}\mathfrak{b})}$$

is given by an element of $\mu_w$, and all these isomorphisms differ at most by (q-1)-st roots of unity.

(5.3) Instead of always calculating mod $\mu_{q-1}$, we consider $\xi(\Lambda)$ and $\theta(\mathfrak{a},\mathfrak{b})$ as elements of $C^*$ resp. $\mu_w$, but keep in mind that multiplicative equations involving $\xi$ and $\theta$ hold true only modulo $\mu_{q-1}$.

5.4 <u>Proposition</u>. Let $\mathfrak{a},\mathfrak{b},\mathfrak{c}$ be integral ideals of $A$. In $C\{\tau\}$, the following relations hold:

(i) $\quad \xi(\mathfrak{a}^{-1}\mathfrak{b}) = \theta(\mathfrak{a},\mathfrak{b})D(\rho_{\mathfrak{a}}^{(\mathfrak{b})})\xi(\mathfrak{b})$ ;

(ii) $\quad \theta(\mathfrak{a}\,\mathfrak{b},\mathfrak{c})\rho_{\mathfrak{a}\mathfrak{b}}^{(\mathfrak{c})} = \theta(\mathfrak{a},\mathfrak{b}^{-1}\mathfrak{c})\rho_{\mathfrak{a}}^{(\mathfrak{b}^{-1}\mathfrak{c})}\theta(\mathfrak{b},\mathfrak{c})\rho_{\mathfrak{b}}^{(\mathfrak{c})}$ ;

(iii) $\quad \theta(\mathfrak{a},\mathfrak{b}) = \mathrm{sgn}(f,\mathfrak{b})$ , if $\mathfrak{a} = (f)$ is principal ;

(iv) $\Theta(\mathfrak{a}\mathfrak{b},\mathfrak{c}) = \Theta(\mathfrak{a},\mathfrak{b}^{-1}\mathfrak{c})\tau^{\deg \mathfrak{a}}(\Theta(\mathfrak{b},\mathfrak{c}))$ ;

(v) $\Theta(\mathfrak{a}\mathfrak{b},\mathfrak{c}) = \mathrm{sgn}(f,\mathfrak{b}^{-1}\mathfrak{c})\Theta(\mathfrak{b},\mathfrak{c})$ , $\mathfrak{a} = (f)$ ;

(vi) $\Theta(\mathfrak{a},\mathfrak{b})$ depends only on the class of $\mathfrak{a}$ in $\widetilde{\mathrm{Pic}}\, A$ and the class of $\mathfrak{b}$ in Pic A .

Proof. (i) and (ii) follow directly from the definitions. By (4.11), for non-zero $f$ in $A$ , we have $\rho_f^{(\mathfrak{b})} = f\tau^\circ + \ldots + \mathrm{sgn}(f,\mathfrak{b})\tau^{\deg f}$ , so $\rho_{(f)}^{(\mathfrak{b})} = \mathrm{sgn}(f,\mathfrak{b})^{-1}\rho_f^{(\mathfrak{b})}$ , and (iii) follows from (i). Assertion (iv) results from (ii) comparing the leading coefficients, (v) is obtained by combining (iii) and (iv), and (vi) is a consequence of (v). □

(5.4 vi) shows that $\Theta(\mathfrak{a},\mathfrak{b})$ may be extended to a map on $\widetilde{\mathrm{Pic}}\, A \times \mathrm{Pic}\, A$ , (iii), (iv) and (v) still holding for fractional ideals.

Putting

$$e_{(\mathfrak{b})} = e_\Lambda$$

for a fractional ideal $\mathfrak{b}$ with $\Lambda = \Lambda^{(\mathfrak{b})}$ , we have by definition for $z \in C$

(5.5) $$\Theta(\mathfrak{a},\mathfrak{b})\rho_{\mathfrak{a}}^{(\mathfrak{b})}(e_{(\mathfrak{b})}(z)) = e_{(\mathfrak{a}^{-1}\mathfrak{b})}(\Theta(\mathfrak{a},\mathfrak{b})D(\rho_{\mathfrak{a}}^{(\mathfrak{b})})z) .$$

5.6 Remarks. (i) By construction, $D(\rho_{\mathfrak{a}}^{(\mathfrak{b})})$ lies in the ring $\tilde{B}$ of integers of $\tilde{H}$ . It generates the ideal $\mathfrak{a}\tilde{B}$ (an explicit version of the principal ideal theorem). By (5.4 ii), it suffices to show this assertion in the case $\mathfrak{a}$ is a prime ideal. But then it is the content of [39, 4.18].

(ii) By means of $D(\rho_{\mathfrak{a}}^{(\mathfrak{b})})$ resp. $e_u(\xi)$ , one may construct units of the rings $\tilde{B}$ resp. $\tilde{B}(\mathfrak{n})$ . The index of such unit groups ("cyclotomic" resp. "elliptic units") in the full group of units is always finite and related to the class number of the corresponding fields. For cyclotomic extensions of $K = \mathbb{F}_q(T)$ , this is carried out in [14,15]. There, one gets results analogous with the unit theorems of Kummer [46] and Sinnott [63]. In [38], Hayes obtained class number formulae for the case $\delta = 1$ and subextensions of $H = \tilde{H}$ . These formulae are similar to those of [56]. Using the results of [39] and of the last section, one should be able to treat the general case $\delta > 1$ as well as the case of ramified extensions.

## V Modular Curves over C

If not explicitly stated, otherwise, all the D-modules in this chapter are assumed of rank 2. G denotes the group scheme GL(2) with center Z .

### 1. The "Upper Half-Plane"

(Essentially, we are giving here a summary of Ch.III in [10]). Let $\Omega = \Omega^2 = \mathbb{P}_1(C) - \mathbb{P}_1(K_\infty) = C - K_\infty$ . On $\Omega$ , $G(K_\infty)$ acts by fractional linear transformations. We define the <u>imaginary absolute value</u> $|z|_i$ of $z \in C$ by

(1.1) $$|z|_i = \inf\{|z-x| \mid x \in K_\infty\} .$$

Trivial properties are:

(i) $K_\infty$ being locally compact, there exists $x \in K_\infty$ with $|z-x| = |z|_i$ ;

(ii) $|z|_i = 0 \iff z \in K_\infty$ ;

(iii) For $c \in K_\infty$ and $z \in C$ , $|cz|_i = |c||z|_i$ ;

(iv) If $|z|$ does not lie in the value group $q^{\delta\mathbb{Z}}$ of $K_\infty^*$ , we have $|z|_i = |z|$ ;

(v) For $z \in C$ with $|z| = 1$ and residue class $\bar{z}$ in the residue field $\bar{k}$ of $C$ , we have $|z|_i = 1 \iff \bar{z} \notin k$ .

Further, for $\begin{pmatrix} a & b \\ c & d \end{pmatrix} = \gamma \in G(K_\infty)$ and $z \in \Omega$ , the equation

(1.2) $$|\gamma z|_i = |\det \gamma||cz+d|^{-2}|z|_i \text{ holds.}$$

<u>Proof</u>. An easy computation shows (1.2) for $\gamma \cdot \gamma'$ , provided it is true for $\gamma$ and $\gamma'$ . Thus it suffices to show

$$|z^{-1}|_i = |z|^{-2}|z|_i ,$$

$G(K_\infty)$ being generated by $\begin{pmatrix} 0 & 1 \\ 1 & 0 \end{pmatrix}$ and the upper triangular matrices, for which the asssertion is trivial. If $|z|_i = z$, we have $|z^{-1}|_i \leq |z^{-1}|$, so $|z^{-1}|_i \leq |z^{-1}| = |z|^{-2}|z|_i$. If $|z|_i = |z-x| < |z|$, then $|z| = |x|$ and $|z^{-1}|_i \leq |z^{-1} - x^{-1}| = |z-x| / |zx| = |z|^{-2}|z|_i$. On the other hand, $|z|_i = |1/z^{-1}|_i \leq |z|^2 |z^{-1}|_i$. □

(1.3) Let $\mathfrak{T}$ be the Bruhat-Tits building of the group $PGL(2,K_\infty)$. It is a connected tree, i.e. a connected, simply connected simplicial complex which has the set of similarity classes of $O_\infty$-lattices in $K_\infty^2$ as its vertex set $\mathfrak{T}(\mathbb{Z})$. (By an $O_\infty$-lattice in $K_\infty^2$, we understand a two-dimensional $O_\infty$-submodule generating $K_\infty^2$.) Two vertices $(L_1) \neq (L_2)$ are adjacent (connected by an edge) if the $L_i$ may be chosen in their classes in such a way that $L_1 \subset L_2$ of index $q^\delta$. Let $L_o$ be the standard lattice $O_\infty^2$. The vertices having distance $n \in \mathbb{N}$ of $(L_o)$ correspond bijectively to the elements of the projective line $\mathbb{P}(L_o/\pi^n L_o)$ over the finite ring $O_\infty/\pi^n$. In particular, each vertex has exactly $q^\delta+1$ neighbors. For a detailed discussion, we refer to [61].

(1.4) The set $\mathfrak{T}(\mathbb{R})$ of points of the realization of $\mathfrak{T}$ may be identified with the set of similarity classes of real-valued norms on the vector space $K_\infty^2$ [25]:

a) To $(L) \in \mathfrak{T}(\mathbb{Z})$, one associates the class of the norm $|\ |_L$ with unit ball $L$:

$$|v|_L = \inf\{|c| \mid c \in K_\infty^*,\ c^{-1}v \in L\} \qquad (v \in K_\infty^2);$$

b) If $x \in \mathfrak{T}(\mathbb{R})$ lies between the adjacent vertices $(L_1)$ and $(L_2)$,

$$x = (1-t)(L_1) + t(L_2),$$

where $0 < t < 1$ and $L_1 \subset L_2$ of index $q^\delta$, $x$ corresponds to the class of the norm $\sup(|\ |_{L_1}, q^{\delta t}|\ |_{L_2})$. (The supremum of two norms, taken elementwise, is again a norm.)

(1.5) Next, we define the building map

$$\lambda : \Omega \longrightarrow \mathfrak{T}(\mathbb{R})$$

$$z \longmapsto \text{class of the norm } |\ |_z,$$

where the value $|(x,y)|_z$ of $|\ |_z$ on $(x,y) \in K_\infty^2$ is given by $|zx+y|$ . By $z \notin K_\infty$ , this defines in fact a norm. Absolute values of elements of $C$ lying in $q^{\mathbb{Q}}$ , $\lambda$ takes values in $\mathfrak{T}(\mathbb{Q})$ and is surjective as a map to $\mathfrak{T}(\mathbb{Q})$ , as is easy to verify.

(1.6) The functions $|z|$ and $|z|_i$ on $\Omega$ factor through $\lambda$ , for we have

$$|z| = \text{length of the vector } (1,0) \text{ w.r.t. } |\ |_z ,$$

$$|z|_i = \text{distance of } (1,0) \text{ to the line } (0,*) \text{ w.r.t. } |\ |_z .$$

Let "log" be the logarithm w.r.t. $q^\delta$ . One easily sees that $\log|z|$ and $\log|z|_i$ are linear functions on $\mathfrak{T}(\mathbb{Q})$ . Now we supply $\mathfrak{T}(\mathbb{R})$ with the metric $d(x,y)$ that gives the distance 1 to adjacent vertices and is linear on edges. With the help of (1.1) and (1.2), it is not hard to verify

(1.7) $$d(\lambda(z),(L_o)) = \begin{cases} -\log|z|_i , & \text{if } |z| \leq 1 , \\ -\log|z^{-1}|_i , & \text{if } |z| \geq 1 . \end{cases}$$

Let now for rational numbers $r \geq 0$

$$\mathfrak{T}(r) = \{x \in \mathfrak{T}(\mathbb{R}) \mid d(x,(L_o)) \leq r\} ,$$

and for $z \in C$

$$B(z,r) = \{y \in C \mid |z-y| < r\} \quad \text{resp.}$$

$$B(\infty,r) = \{y \in C \mid |y| > r^{-1}\} \cup \{\infty\}$$

the "open" balls in $\mathbb{P}_1(C)$ with radius $r$ . (There should be no confusion between the place $\infty$ of $K$ and the point $\infty$ in $\mathbb{P}_1(C)$ !)

From (1.7) and (1.1v) we get

$$\lambda^{-1}((L_o)) = \lambda^{-1}(\mathfrak{T}(0)) = \{z \in C \mid |z| = |z|_i = 1\}$$

$$= \mathbb{P}_1(C) - (B(0,1) \cup B(\infty,1) \cup \bigcup_{x \in k^*} B(x,1)) .$$

Correspondingly, for $0 \leq r < 1$

$$\lambda^{-1}(\mathfrak{T}(r)) = \mathbb{P}_1(C) - \bigcup_{x \in \mathbb{P}_1(k)} B(x,s) ,$$

where $s = q^{-r\delta}$ .

Increasing $r$ , the radius of the balls $B(x,s)$ decreases, each of these balls splitting into a disjoint union of $q^{\delta}$ balls of radius $q^{-\delta}$ if $r$ takes the limit value $r = 1$ . For $r$ general, one obtains [10, Ch.III 5.4]:

(1.8) Let $n \geq 0$ be an integer and $n \leq r < n+1$ . Then $\lambda^{-1}(\mathfrak{T}(r))$ is the complement in $\mathbb{P}_1(C)$ of a finite number of disjoint balls $B(x,s)$ . The radii $s$ are given by $s = q^{-r\delta}$ , and the set of these balls is in bijection with the set $\mathbb{P}_1(O_\infty/\pi^{n+1})$ .

(1.9) The sets $\lambda^{-1}(\mathfrak{T}(r))$ are connected affinoid subdomains of $\mathbb{P}_1(C)$ ; their union $\Omega$ with its induced analytic structure is an unbounded Stein domain [13,24]. For example, this implies $\mathrm{Aut}\,\Omega = PGL(2,K_\infty)$ for the group of analytic automorphisms of $\Omega$ [54, 1.6] . However, in contrast with the complex upper half-plane, $\Omega$ is not simply connected.

(1.10) An intuitive topological picture of $\lambda : \Omega \to \mathfrak{T}(\mathbb{R})$ may be obtained as follows: For each vertex of $\mathfrak{T}$ , take one copy of

$$\mathbb{P}_1(\mathbb{C}) - \begin{pmatrix}\text{union of } q^{\delta}+1 \text{ open balls} \\ \text{whose closures are all disjoint}\end{pmatrix} ,$$

for each edge of $\mathfrak{T}$ , take an annulus

$$\mathbb{P}_1(\mathbb{C}) - (\text{union of } 2 \text{ disjoint open balls}) ,$$

and glue corresponding to the incidence relations on $\mathfrak{T}$ . The resulting two-dimensional manifold is the boundary of a tubular neighborhood of $\mathfrak{T}(\mathbb{R})$ , and $\lambda$ corresponds to the projection onto $\mathfrak{T}(\mathbb{R})$ .

## 2. Group Actions

The group $G(K_\infty)$ operates as a group of matrices from the right on $K_\infty^2$ and from the left on the set of norms, i.e. on $\mathfrak{T}(\mathbb{R})$ . For $\gamma \in G(K_\infty)$ , $v \in K_\infty^2$ , and $N$ a norm on $K_\infty^2$ , $(\gamma N)(v) = N(v\gamma)$ . Clearly:

(2.1) (i) $G(K_\infty)$ acts simplicially on $\mathfrak{T}(\mathbb{R})$ ;

(ii) $G_+(K_\infty) = \{\gamma \in G(K_\infty) \mid \det\gamma$ has an even valuation$\}$ acts orientation-preserving;

(iii) the building map $\lambda$ is $G(K_\infty)$-equivariant.

(2.2) An <u>arithmetic subgroup</u> $\Gamma$ of $GL(2,K)$ is a congruence subgroup of the group $GL(Y)$ of a 2-lattice $Y \subset K^2$ , i.e. a subgroup of $GL(Y)$ containing the kernel $GL(Y,\mathfrak{n})$ of $GL(Y) \to GL(Y/\mathfrak{n}Y)$ for some ideal $\mathfrak{n} \subset A$ . A group containing no elements of finite order prime to $p$ is called <u>p'-torsion free</u>.

An arithmetic subgroup $\Gamma$ acts orientation-preserving and with finite stabilizers on $\mathfrak{T}$ , and $\Gamma\backslash\mathfrak{T}$ is still a graph, i.e. a one-dimensional cell complex. In general, $\Gamma\backslash\mathfrak{T}$ is not a simplicial complex for it may happen that adjacent vertices are connected by different edges. The building map induces

$$\lambda_\Gamma : \Gamma\backslash\Omega \to \Gamma\backslash\mathfrak{T}(\mathbb{R}) ,$$

and both analytic structures on $\Gamma\backslash\Omega$ (as a quotient of $\Omega$ , and as the set of C-valued points of the affine algebraic curve $M_\Gamma$ (II 1.8)) coincide [11, 6.6].

Let us first describe the canonical nonsingular compactification $\bar{M}_\Gamma$ of $M_\Gamma$ , i.e.

a) the set $\bar{M}_\Gamma - M_\Gamma$ of <u>cusps</u> of $\Gamma$ ;

b) the analytic structure of $\bar{M}_\Gamma$ around a cusp.

For the moment, we do not care about the field of definition of $M_\Gamma$ . By "points", we understand C-valued points, and we simply write $M_\Gamma$ instead of $M_\Gamma(C)$ etc.

(2.3) The graph $\Gamma\backslash\mathfrak{T}$ is the union of a finite graph $(\Gamma\backslash\mathfrak{T})^o$ and a finite number of ends. An <u>end</u> is an infinite graph of type
* —— * —— * —— *..... Two ends are <u>equivalent</u> if they differ at most by a finite graph.

The equivalence classes of ends occurring in $\Gamma\backslash\mathfrak{T}$ are in canonical 1-1 correspondence with $\Gamma\backslash\mathbb{P}_1(K)$ . Namely,

$$\{\text{classes of ends of } \mathbb{C}\} = \varprojlim_{n \in \mathbb{N}} \mathbb{P}(L_o/\pi^n L_o) = \mathbb{P}_1(O_\infty) = \mathbb{P}_1(K_\infty) ,$$

and precisely the parabolic fixed points of $\Gamma$, i.e. the elements of $\mathbb{P}_1(K)$ give rise to ends of $\Gamma\backslash\mathbb{C}$. (For all these assertions, see [61, Ch.II].)

(2.4) Combining (1.8) and (2.3), we get the following alternative descriptions for the set $Sp(\Gamma)$ of cusps of $\Gamma$:

a) $\bar{M}_\Gamma - M_\Gamma$,

b) {classes of ends in $\Gamma\backslash\mathbb{C}$},

c) $\Gamma\backslash\mathbb{P}_1(K)$.

If $\Gamma$ equals the full group $GL(Y)$ of a lattice $Y$, an element $s$ of $\mathbb{P}_1(K)$ defines a short exact sequence

$$0 \to U(Y,s) \to Y \to V(Y,s) \to 0$$

with projective A-modules of rank 1. Elementary lattice theory [3, VII, § 4, 10] implies the bijectivity of the map

$$\Gamma\backslash\mathbb{P}_1(K) \longrightarrow \mathrm{Pic}\, A$$

$$s \longmapsto \text{class of } V(Y,s) .$$

In this case, the number of cusps is given by $\delta \cdot h = \delta \cdot P(1)$ with the numerator polynomial $P(X)$ of $Z_K$.

(2.5) For each element of $\Gamma\backslash\mathbb{P}_1(K)$, we have to adjoin a point to $M_\Gamma = \Gamma\backslash\Omega$ and to describe a local parameter. So let $s \in \mathbb{P}_1(K)$ and $\nu \in G(K)$ with $\nu(\infty) = s$. For a meromorphic function $f$ on $\Gamma\backslash\Omega$, i.e. a $\Gamma$-invariant function on $\Omega$, the function $f_\nu = f \circ \nu$ is invariant by $\Gamma^{\nu^{-1}}$, and we will describe the behavior of $f$ at $s$ by that of $f_\nu$ at $\infty$. The stabilizer $(\Gamma^{\nu^{-1}})_\infty = (\Gamma_s)^{\nu^{-1}}$ contains a maximal subgroup of the form $\left\{\begin{pmatrix}1 & b\\ 0 & 1\end{pmatrix} \middle| b \in \mathfrak{b}\right\}$ with some fractional ideal $\mathfrak{b} = \mathfrak{b}(\nu,\Gamma)$ of $A$. Like the lattice function $e_\mathfrak{b}$, $f_\nu$ is invariant by the translations $z \longmapsto z+b$ $(b \in \mathfrak{b})$. There is an admissible $\mathfrak{b}$-stable subset $\Omega'$ of $\Omega$ such that $e_\mathfrak{b}^{-1}$ identifies $\mathfrak{b}\backslash\Omega'$ with a pointed neighborhood of zero in $C$. (For instance, one may use $\Omega' = \{z \in \Omega \,|\, |z|_i \geq r\}$, where $r \gg 0$ [17, 3.2.17] or [27, 1.76].)

(2.6) Put $t = t(\nu,\Gamma) = e_{\mathfrak{h}}^{-1}$ .

If $\Gamma_s$ is p'-torsion free (for example, if $\Gamma$ is a full congruence subgroup of $GL(Y)$ , we use $t$ as a parameter.

In general, $(\Gamma\nu^{-1})_\infty$ will be of the form $\left\{\begin{pmatrix} a & b \\ 0 & d \end{pmatrix}\right\}$ with $b \in \mathfrak{h}$ and certain $a,d \in \mathbb{F}_q^*$ , i.e. it will contain transformations of type $z \longmapsto az$ . If $w$ is the order of the cyclic group of these transformations, the correct parameter is $t^w$ . Of course, $w = q-1$ in case $\Gamma = GL(Y)$ .

(2.7) Thus, the function $f$ is holomorphic at the cusp $s$ if $f_\nu$ has an expansion

$$f_\nu(t) = \sum_{i \geq 0} a_i t^{wi}$$

with a positive radius of convergence. Following the same lines, one defines meromorphic functions, the order of zero at $s$ , etc. .

2.8 Remarks. (i) In order to describe the behavior of $f$ , we have chosen: a) $s$ in its class mod $\Gamma$ , b) $\nu \in G(K)$ with $\nu(\infty) = s$ . As one easily verifies, the fact of being holomorphic or meromorphic, and the order of zero of $f$ , does not depend on these choices, but the coefficients $a_i$ do. Thus, it makes no sense to speak of "the" expansion of $f$ at a cusp.

(ii) Let $\Gamma$ be an arithmetic subgroup of $G(K)$ . The equation $\begin{pmatrix} a & b \\ 0 & a \end{pmatrix}^q = \begin{pmatrix} a & 0 \\ 0 & a \end{pmatrix}^q$ forces $\Gamma_\infty$ to contain $\begin{pmatrix} a & 0 \\ 0 & a \end{pmatrix}$ whenever it contains $\begin{pmatrix} a & b \\ 0 & a \end{pmatrix}$ . Therefore, the phenomenon of irregular cusps [62, p.29] cannot occur.

(iii) We identify the three sets in (2.4) and denote them by $Sp(\Gamma)$ . Sometimes, elements of $\mathbb{P}_1(K)$ will called "cusps", but the context will always show whether elements of $\mathbb{P}_1(K)$ or $\Gamma$-equivalence classes are meant.

## 3. Modular Forms

3.1 Definition. A holomorphic modular form of weight $k$ for the arithmetic group $\Gamma$ is a $C$-valued function $f$ on $\Omega$ having the following properties:

(i) For $\gamma = \begin{pmatrix} a & b \\ c & d \end{pmatrix}$ in $\Gamma$ and $z$ in $\Omega$, we have $f(\gamma z) = (cz+d)^k f(z)$ ;

(ii) $f$ is holomorphic on $\Omega$ ;

(iii) $f$ is holomorphic at the cusps of $\Gamma$ .

Condition (iii) needs some explanation. Define for $\gamma \in G(K)$ and a function $f$ on $\Omega$

$$f_{[\gamma]}(z) = f_{[\gamma]_k}(z) = (cz+d)^{-k} f(\gamma z) \quad .$$

This defines a right operation, i.e. $f_{[\gamma\gamma']} = f_{[\gamma][\gamma']}$ . By (i), i.e. $f_{[\gamma]} = f$ , we get $f_{[\nu][\gamma']} = f_{[\nu]}$ , where $\gamma \in \Gamma$ , $\nu$ as in (2.5), $\gamma' = \gamma^{\nu^{-1}}$ . This says $f_{[\nu]}$ is invariant by translations $z \longmapsto z+b$ , $b \in \mathfrak{b}(\nu,\Gamma)$ . Now (iii) means: $f_{[\nu]}$ has a convergent series expansion with respect to $t(\nu,\Gamma)$ . In the same way, one defines meromorphic modular forms, zero orders etc. . Remark (2.8i) will hold accordingly.

(3.2) We put $M_k(\Gamma)$ (resp. $S_k(\Gamma)$ ) for the $C$-vector space of holomorphic modular forms (resp. cusp forms = holomorphic modular forms that vanish at all the cusps) of weight $k$ .

The group $\Gamma = GL(Y)$ contains $Z(\mathbb{F}_q)$ as its center. As a consequence of (i), $M_k(\Gamma) = 0$ if $k \not\equiv 0(q-1)$ .

Let $Y_\omega \subset C$ be the image of $Y$ under the imbedding $i_\omega : K^2 \hookrightarrow C$ , defined by the values $\omega$ resp. 1 of the standard basis vectors (1,0) resp. (0,1) .

3.3 Example. The Eisenstein series $E^{(k)}(\omega) = E^{(k)}(Y_\omega) = \sum'_{\lambda \in Y_\omega} \lambda^{-k}$ ,

where $k \in \mathbb{N}$ , $k \equiv 0(q-1)$ , is a holomorphic modular form of weight $k$ for $\Gamma = GL(Y)$ . The transformation behavior under $\Gamma$ and the holomorphy on $\Omega$ may be verified directly. The holomorphy at cusps will follow for example from (VI 3.9), see also [27].

3.4 Example. Let $\phi$ be the D-module associated with $Y_\omega$. For $a \neq 0$, write

$$\phi_a = \sum_{i \leq 2 \deg a} l_i(a,\omega)\tau^i$$

and consider $l_i$ as a function on $\Omega$. By (II 3.2), $l_i(\gamma\omega) = (cz+d)^k l_i(\omega)$ for $\gamma \in \Gamma = GL(Y)$ and $k = q^i-1$. The $l_i$ are modular forms of weight $q^i-1$ for $\Gamma$. They are related with the $E^{(k)}$ by (II 2.11). We may still generalize this construction. Instead of $\phi_a = \phi_a^{Y_\omega}$, let us consider $\phi_{\mathfrak{a}}^{Y_\omega}$, the isogeny corresponding to the inclusion of lattices $Y_\omega \hookrightarrow \mathfrak{a}^{-1}Y_\omega$ (II 3.3, 3.4), where $\mathfrak{a} \subset A$ is an ideal, and $\phi_{\mathfrak{a}}^{Y_\omega}$ is normalized by $D(\phi_{\mathfrak{a}}^{Y_\omega}) = 1$. Again, the coefficients $l_i(\phi_{\mathfrak{a}}^{Y_\omega})$ of $\tau^i$ are holomorphic modular forms of weight $q^i-1$ for $GL(Y)$. We define

$$\Delta_a = l(\phi_a) = l_{2\deg a}(\phi_a)$$

$$\Delta_{\mathfrak{a}} = l(\phi_{\mathfrak{a}}) = l_{2\deg \mathfrak{a}}(\phi_{\mathfrak{a}}) .$$

Note however: For the principal ideal $(a)$, we have $\phi_a = a\tau^o + \ldots \Delta_a\tau^{2\deg a}$, but $\phi_{(a)} = \tau^o + \ldots \Delta_{(a)}\tau^{2\deg a}$, i.e. $\Delta_a = a\Delta_{(a)}$.

3.5 Remark. There is a certain inconsistency in notation. In (IV 5.1), the isogeny $\rho_{\mathfrak{a}}$ induced by an ideal $\mathfrak{a}$ on a 1-D-module $\rho$ has been normalized by $l(\rho_{\mathfrak{a}}) = 1$, in contrast with the normalization $D(\phi_{\mathfrak{a}}) = 1$ presently used. Accordingly, $\Delta_{\mathfrak{a}}^{-1}$ is a two-dimensional analogue of $D(\rho_{\mathfrak{a}})$.

3.6 Example [28]. Let $(K,A,\infty) = (\mathbb{F}_q(T), \mathbb{F}_q[T], \infty)$, $Y = A^2$, $\Gamma = GL(2,A)$, $\phi_T^{Y_\omega} = T\tau^o + g(\omega)\tau + \Delta(\omega)\tau^2$. Then $g \in M_{q-1}(\Gamma)$, $\Delta \in S_{q^2-1}(\Gamma)$, and

$$C[g,\Delta] = \bigoplus_{k \geq 0} M_k(\Gamma) .$$

The j-invariant $j = g^{q+1}/\Delta$ identifies $\Gamma\backslash\Omega$ with the affine j-line over $C$. But in general, the situation is much more complicated.

Next, we construct modular forms for the subgroups $\Gamma(\mathfrak{n}) = GL(Y,\mathfrak{n})$ of $\Gamma$, $\mathfrak{n}$ being an integral ideal.

3.7 Example. Let $u = (u_1,u_2) \in \mathfrak{n}^{-1}Y-Y$. We put

$$e_u(\omega) = e_{Y_\omega}(u_1\omega+u_2) ,$$

and for $k \in \mathbb{N}$

$$E_u^{(k)}(\omega) = \sum_{\substack{v \in K^2 \\ v \equiv u \bmod Y}} i_\omega(v)^{-k} .$$

The $E_u^{(k)}(\omega)$ lie in $M_k(\Gamma(\mathfrak{n}))$ . Again, only the holomorphy at cusps is non-trivial; it will follow from (VI 3.9). For $\gamma \in \Gamma$ ,

(i) $\quad E_u^{(k)}(\gamma\omega) = (c\omega+d)^k E_{u\gamma}^{(k)}(\omega)$

holds. Further, $u_1\omega+u_2 \in \mathfrak{n}^{-1}Y_\omega$ implies

(ii) $\quad \phi_n^{Y_\omega}(e_u(\omega)) = e_{Y_\omega}(n(u_1\omega+u_2)) = 0$

for each $n \in \mathfrak{n}$ . Some other consequences of the properties of e-functions are

(iii) $\quad e_u$ depends only on $u \bmod Y$ and has no zeroes on $\Omega$ ;

(iv) $\quad e_u^{-1}(\omega) = E_u^{(1)}(\omega)$ $\qquad$ (I 2.2v) .

$e_u$ is a meromorphic modular form of weight $-1$ for $\Gamma(\mathfrak{n})$ which has poles of (generally) strictly positive orders at cusps.

The forms of (3.3) and (3.4) are derived from lattice functions of weight $k$ which are naturally defined on all components. Thus we make the following

3.8 Definition. Consider the set of pairs $(\Lambda,\alpha)$ , where $\Lambda$ is a 2-lattice in $C$ with a $\mathfrak{n}$-structure $\alpha$ . A modular form of weight $k$ and level $\mathfrak{n}$ is a map from this set to $C$ satisfying

$$f(c^{-1}\Lambda, c\alpha) = c^k f(\Lambda,\alpha) \qquad (c \in C^*) ,$$

as well as the holomorphy properties corresponding to (3.1). In the same way, we may define meromorphic modular forms.

A modular form $f$ of level $\mathfrak{n}$ is the same as a family $\{f_{Y,\alpha}\}$ of modular forms $f_{Y,\alpha}$ for $GL(Y,\mathfrak{n})$ , where $Y$ runs through a RS of $\mathfrak{P}_A^2$ and $\alpha$ through a RS of $\mathfrak{n}$-structure on $Y$ .

## 4. Elliptic Points

Let again $\Gamma = GL(Y)$ . The group of elements of $\Gamma$ operating trivial on $\Omega$ consists of the scalars in $\Gamma$ , i.e. agrees with the group $Z(\mathbb{F}_q)$ .

4.1 Definition. A point $\omega \in \Omega$ is called an elliptic point of $\Gamma$ if the stabilizer $\Gamma_\omega$ is strictly larger than $Z(\mathbb{F}_q)$ . Let $E$ be the set of elliptic points and $Ell(\Gamma) = \Gamma \setminus E$ . Accordingly, the elements of $Ell(\Gamma) \subset \Gamma \setminus \Omega$ are sometimes called elliptic points.

Elliptic points $\omega$ satsify a quadratic equation $c\omega^2 + (d-a)\omega + b = 0$ , where $\gamma = \begin{pmatrix} a & b \\ c & d \end{pmatrix} \in \Gamma$ , $c \neq 0$ . The element $\gamma$ is of finite order prime to $p$ , $\lambda(\omega)$ being a fixed point of $\gamma$ on $\mathbb{C}(\mathbb{R})$ . (The elements in $G(K)$ of order $p$ are conjugate in $G(K)$ to strictly upper triangular matrices that fix no element of $\Omega$ .) Consequently, the eigenvalues $\varepsilon_1, \varepsilon_2$ of $\gamma$ are roots of unity, and $\varepsilon_1 \neq \varepsilon_2$, $\varepsilon_i \notin \mathbb{F}_q$ . Hence, the elliptic point $\omega$ satisfies

(4.2) $$K(\omega) = K \cdot \mathbb{F}_{q^2}$$

with the quadratic constant field extension $\mathbb{F}_{q^2}$ .

An elliptic $\omega$ cannot lie in $K_\infty$ ; so the existence of elliptic points implies $\mathbb{F}_{q^2} \not\hookrightarrow K_\infty$ , i.e. $\delta$ is odd. Therefore, we assume $\delta \equiv 1(2)$ for the rest of this section. Let $K' = K \cdot \mathbb{F}_{q^2} \subset C$ and $A' = A\,\mathbb{F}_{q^2}$ = integral closure of $A$ in $K'$ . Then $Ell(\Gamma)$ corresponds to the set $\{\mathfrak{a}'\}/K'^*$ , where $\mathfrak{a}'$ runs through the set of lattices of rank 2 in $K' \subset C$ satisfying

a) $Aut(\mathfrak{a}') \supsetneq \mathbb{F}_q^*$ and

b) $\mathfrak{a}'$ is isomorphic with $Y$ as an A-module.

Let $\mathfrak{a}'$ be a lattice of rank 2 in $K'$ . We have the following well-known facts (see [62, IV 4,5.4.2] or [36, § 1]):

(4.3) (i) $End(\mathfrak{a}')$ is an order $\tilde{A} \subset A'$ , i.e. $\tilde{A}$ is of finite index and contains $A$ ;

(ii) $\tilde{A} = A + \mathfrak{n}A'$ with an ideal $\mathfrak{n}$ of $A$ ;

(iii) $Aut(\mathfrak{a}') = \tilde{A}^*$ .

The set $\{a'\}$ decomposes according to the different orders $\tilde{A}$ in $A'$, and

(iv) $a'$ belongs to $\tilde{A} \Leftrightarrow a' = \tilde{A}\underline{x}$ with some $\underline{x} \in I_{f,K'}$, i.e. for each place $\mathfrak{p}$ of $A$, there exists $\underline{x}_{\mathfrak{p}} \in K' \otimes_A A_{\mathfrak{p}}$ such that $a' \otimes A_{\mathfrak{p}} = \tilde{A} \otimes A_{\mathfrak{p}} \cdot \underline{x}_{\mathfrak{p}}$.

The assumption $\mathbb{F}_q^* \underset{\neq}{\subset} A^*$ implies $\mathbb{F}_{q^2} \subset \tilde{A}$, i.e. $\tilde{A} = A'$. Therefore, $\mathrm{Ell}(\Gamma)$ corresponds precisely to the set of classes of ideals $a'$ of $A'$ which are isomorphic with $Y$ as an A-module. By (II 1.4) and the unramifiedness of $A'$ above $A$, this condition is equivalent with the following: The norm $N_K^{K'}(a')$ and the second exterior power $\Lambda^2(Y)$ determine the same ideal class.

Now consider the commutative diagram

$$
(4.4)\qquad
\begin{array}{ccccccccc}
0 & \longrightarrow & J(\mathbb{F}_{q^2}) & \longrightarrow & \mathrm{Pic}\,A' & \xrightarrow{\deg} & \mathbb{Z}/\delta & \longrightarrow & 0 \\
 & & \Big\downarrow{\scriptstyle \mathrm{norm}} & & \Big\downarrow{\scriptstyle \mathrm{norm}} & & \Big\downarrow{\scriptstyle 2} & & \\
0 & \longrightarrow & J(\mathbb{F}_q) & \longrightarrow & \mathrm{Pic}\,A & \longrightarrow & \mathbb{Z}/\delta & \longrightarrow & 0\,.
\end{array}
$$

The norm kernels in $J(\mathbb{F}_{q^2})$ and in $\mathrm{Pic}\,A'$ agree, $\delta$ being odd. The surjectivity of the norm on $J$ gives the value $\#(J(\mathbb{F}_{q^2}))/\#(J(\mathbb{F}_q))$ for the number $\#(\mathrm{Ell}(\Gamma))$ of elliptic classes. As is easy to see, this number agrees with the value $P(-1)$ of the numerator polynomial $P(X)$ of (III 1.6).

All the relevant facts are collected in the

4.5 Proposition. Let the degree $\delta$ of the place $\infty$ be even. Then $\Gamma = GL(Y)$ possesses no elliptic points. If $\delta$ is odd, there are precisely $P(-1)$ $\Gamma$-classes of elliptic points, each having a $\Gamma$-stabilizer isomorphic with $\mathbb{F}_{q^2}^*$.

## 5. Modular Forms and Differentials

As in the classical case, there is a connection between differentials on the curve $\overline{M}_\Gamma$ and modular forms for $\Gamma$. Using Kiehl's theorems of GAGA-type [43,44], we need not distinguish between "algebraic" and "analytic" differentials. Since

$$
(5.1)\qquad \frac{d}{dz}(\gamma z) = \det\gamma\,(cz+d)^{-2} \qquad \left(\gamma = \begin{pmatrix} a & b \\ c & d \end{pmatrix}\right),
$$

the differential $dz$ on $\Omega$ is formally of weight $-2$, at least for the group $SL(Y)$.

(5.2) Let $e \in \Omega$ be an elliptic point of $\Gamma = GL(Y)$, and let

$$y = (z-e)^{q+1}u(z) \qquad (u(e) \neq 0)$$

be a local parameter for $\Gamma\backslash\Omega$ around $e$. For a natural number $k$ satisfying $k \equiv 0(q+1)$ as well as $k \equiv 0(q-1)$ and a holomorphic modular form $f(z)$ for $\Gamma$ of weight $2k$, $f(z)(dz)^k$ is an invariant differential form of degree $k$ which may be developed with respect to $y$. By

$$dy = (z-e)^q(u(z)+u'(z)(z-e))dz ,$$

in a neighborhood of $y = 0$,

$$f(z)(dz)^k = F(y)y^{-kq/(q+1)}(dy)^k$$

with some holomorphic function $F(y)$ satisfying $F(0) = f(e)$. Hence, $f(z)(dz)^k$ has a pole of order $\leq kq/(q+1)$ at $y = 0$.

(5.3) Next, we consider the situation at cusps of $\Gamma$. For a cusp $s$, we have (notations as in (2.5, 2.6)):

$$(\Gamma^{\nu^{-1}})_\infty = \left\{\begin{pmatrix} a & b \\ 0 & d \end{pmatrix} \middle| a,d \in \mathbb{F}_q^* , b \in \mathfrak{b}\right\} .$$

Let $t = t(\nu,\Gamma)$ and $y = t^{q-1}$. According to (5.2), we want to express $f(z)(dz)^k$ by means of $y$. By

$$\frac{dy}{dz} = \frac{d}{dz} e_{\mathfrak{b}}^{1-q}(z) = e_{\mathfrak{b}}^{-q}(z) ,$$

$$(dy)^k = y^{kq/(q-1)}(dz)^k .$$

Let us first assume $s = \infty$, $\nu = 1$. Then for $F(y) = f(z)$,

$$f(z)(dz)^k = F(y)y^{-kq/(q-1)}(dy)^k$$

which has a pole of order $\leq kq/(q-1)$ at $y = 0$. If $\nu = \begin{pmatrix} a & b \\ c & d \end{pmatrix}$ not necessarily equals 1, the same is true for $F(y) = f_{[\nu]_{2k}}(z)$ and the

$\Gamma^{\nu^{-1}}$ - invariant differential form

$$f(\nu z)(d(\nu z))^k = f_{[\nu]_{2k}}(z)(dz)^k$$
$$= F(y)y^{-kq/(q-1)}(dy)^k .$$

Away from elliptic points and cusps, there are no problems to identify germs of invariant differentials and germs of modular forms. This shows:

5.4 Theorem. Let $\Gamma = GL(Y)$ and $k$ a natural number divisible by $(q-1)$ and $(q+1)$. Then the $\bar{M}_\Gamma$-sheaf $\mathfrak{m}_{2k}$ of germs of holomorphic modular forms of weight $2k$ is isomorphic with the sheaf of differential forms of degree $k$ with poles at most at the cusps (of order $\leq kq/(q-1)$) and at the elliptic points (of order $\leq kq/(q+1)$).

5.5 Corollary. If $0 \neq f \in M_{2k}(\Gamma)$ has a divisor of degree $d$ ( $k$ as in (5.4)), then

$$d = k[2g(\bar{M}_\Gamma)-2+qs/(q-1)+qe/(q+1)] ,$$

where $g(\bar{M}_\Gamma)$ denotes the genus of $\bar{M}_\Gamma$, and $s$ (resp. $e$) is the number of cusps (resp. of elliptic points) of $\Gamma$ given by (2.4) (resp. (4.5)).

After having the expansions of the modular forms $\Delta_a$ around cusps at our disposal, we will use (5.5) to compute $g(\bar{M}_\Gamma)$.

(5.6) The subgroups $\Gamma(\mathfrak{n}) \subset \Gamma$ for $\mathfrak{n} \subsetneq A$ are p'-torsion free, so they have no elliptic points. Further, the $t(\nu,\Gamma(\mathfrak{n}))$ themselves are parameters at the cusps of $\Gamma(\mathfrak{n})$. Carrying out the analogous considerations, we obtain

5.7 Theorem (compare [27, 1.80]). Let $k \in \mathbb{N}$ and $\mathfrak{n} \subsetneq A$. The $\bar{M}_{\Gamma(\mathfrak{n})}$-sheaf of germs of holomorphic modular forms of weight $2k$ is isomorphic with the sheaf of differential forms of degree $k$ of $\bar{M}_{\Gamma(\mathfrak{n})}$ with poles of order $\leq 2k$ at the cusps (and no further poles).

The maximal pole order comes out by

$$dt/dz = de_{\mathfrak{b}}^{-1}(z)/dz = -e_{\mathfrak{b}}^{-2}(z) = -t^2 .$$

## Appendix: The First Betti Number of $\Gamma$

The results of this section will not further be used in this work. We discuss the relations between the cohomology of an arithmetic subgroup $\Gamma$ of $G(K)$ and the l-adic cohomology of the curve $\bar{M}_\Gamma$. Comparing with the computations of Chapter VI, this will give an answer to a question left open in [61]. The result (A.11) as well as the idea of proof is contained implicitly in [11, § 10], where $H^1(\bar{M}^2, \mathbb{Q}_l)$ is interpreted as a space of automorphic forms. A similar presentation may be found in [10, Ch. IV). Since this text has not been published, we will give a sketch of the argument and the remarks necessary to derive (A.11) from [11].

Let $G$ be a graph (which means here: a locally finite, one-dimensional, oriented cell complex) having the sets $E(G)$, $K(G)$, $P(G)$ of vertices, edges, arrows (= oriented edges).

A.1 Definition. A mapping $c$ of $P(G)$ into an abelian group $U$ will be called

(i) alternating, if for each $\vec{k} \in P(G)$, we have $c(\vec{k}) = -c(\overleftarrow{k})$,

(ii) harmonic, if for each $e \in E(G)$, we have

$$\sum_{\vec{k} \to e} c(\vec{k}) = 0 .$$

Here, $\overleftarrow{k}$ denotes the arrow $\vec{k}$ with inverse orientation, and the sum in (ii) runs over the arrows with head $e$.

The group of harmonic, alternating mappings of $P(G)$ into $U$ is denoted by $\underline{H}^1(G,U)$.

By means of $\lambda_\Gamma$, we are going to compare the cohomology of $\bar{M}_\Gamma$ with that of $\Gamma\backslash\mathfrak{T}$. For this purpose, we introduce an ad-hoc version of a rigid analytic étale cohomology theory. In the following, $n$ is a natural number prime to $p$, and $W$ an analytic space over $C$ with structural sheaf $\mathcal{O}$.

(A.2) We define $H^o(W,\mathbb{Z}/n)$ and $H^o(W,\mu_n)$ as the corresponding group of sections in the rigid analytic category, $\mathbb{Z}/n$ being the constant sheaf, and $\mu_n$ the sheaf kernel of $\mathcal{O}^* \xrightarrow{n} \mathcal{O}^*$ .

Let further $H^1(W,\mu_n)$ be the group of pairs $(\mathcal{L},\beta)$ up to isomorphism, where $\mathcal{L}$ is an invertible $\mathcal{O}$-sheaf and $\beta$ an isomorphism $\mathcal{O} \xrightarrow{\cong} \mathcal{L}^{\otimes n}$ , the group structure being induced from the tensor product. Note that all these notions make sense for rigid analytic spaces. Finally, put $H^1(W,\mathbb{Z}/n) = H^1(W,\mu_n) \otimes \mu_n^{-1}$ .

The following properties are stated without proof:

(A.3) If $W = V_{an}$ is the analytic space associated with a projective algebraic variety $V$ over $C$ , for $i = 0,1$ and the coefficient sheafs $\mu_n$ and $\mathbb{Z}/n$ , we have $H^i(W) = H^i(V)$ . The assertion remains true for $V$ an affine algebraic curve. (Here, $H^i(V)$ denotes the étale cohomology of algebraic varieties.)

(A.4) Let $\{W_i\}$ be an admissible open covering of $W$ with a nerve of dimension $\leq 1$ . There is an exact sequence of cohomology groups with values in $\mu_n$ :

$$0 \to H^o(W) \to \prod_i H^o(W_i) \to \prod_{i \neq j} H^o(W_i \cap W_j) \to$$

$$\to H^1(W) \to \prod_i H^1(W_i) \to \prod_{i \neq j} H^1(W_i \cap W_j) .$$

(A.5) If $f : W \to W'$ is a finite étale morphism with Galois group $G$ having an order prime to $n$ , $H^1(W',\mu_n)$ is canonically isomorphic with the group $H^1(W,\mu_n)^G$ of G-invariants.

The only assertion lying deeper is (A.3) which follows from the GAGA-type theorems of [43,44].

Our first step is the computation of the cohomology of rational domains in $\mathbb{P}_1(C)$ .

A.6 <u>Proposition</u>. Let $B_o \ldots B_m$ be $m+1$ pairwise disjoint open balls in $\mathbb{P}_1(C)$ whose radii lie in $q^{\mathbb{Q}}$ , and let $W = \mathbb{P}_1(C) - \cup B_i$ . Then

(i) $H^o(W,\mathbb{Z}/n) = \mathbb{Z}/n$ and

(ii) $H^1(W,\mathbb{Z}/n) = (\mathbb{Z}/n)^m$ .

For the proof, see [11, 10.1] or [13, V3]. (i) says that $W$ is connected, and for (ii), one has to compute the cohomology of an annulus and to relate the different annuli by a "Kummer sequence". Note that (A.6 ii) would not be true for the ordinary cohomology of rigid analytic spaces.

A.7 Proposition. We have $H^0(\Omega,\mathbb{Z}/n) = \mathbb{Z}/n$, and there is a $\Gamma$-equivariant isomorphism

$$H^1(\Omega,\mu_n) \xrightarrow{\cong} \underline{H}^1(\mathfrak{T},\mathbb{Z}/n) .$$

Proof (sketch). The first assertion holds since $\Omega$ is connected. Let for the moment $I$ be the index set $E(\mathfrak{T}) \cup K(\mathfrak{T})$. We consider the covering $\Omega = \bigcup_{i \in I} \Omega_i$, where we set

$$\Omega_e = \{\omega | d(\lambda(\omega),e) \leq 1/3\} \quad \text{resp.}$$

$$\Omega_k = \{\omega | \lambda(\omega) \in k,\ d(\lambda(\omega),\ E(\mathfrak{T})) \geq 1/4\}$$

for $e \in E(\mathfrak{T})$ resp. for $k \in K(\mathfrak{T})$.

If $i$ and $j$ are not incident then $\Omega_i \cap \Omega_j = \emptyset$; if, however, $e$ is a vertex of the edge $k$ then $\Omega_e \cap \Omega_k$ is an annulus, and $H^1(\Omega_k,\mu_n) \xrightarrow{\cong} H^1(\Omega_e \cap \Omega_k,\mu_n)$. By (A.4), this implies the injectivity of the canonical maps

$$H^1(\Omega,\mu_n) \to \prod_{i \in I} H^1(\Omega_i,\mu_n)$$

and even of

$$H^1(\Omega,\mu_n) \to \prod_{k \in K(\mathfrak{T})} H^1(\Omega_k,\mu_n) .$$

An orientation of the edge $k$ fixes an orientation of the annulus $\Omega_k$ and an isomorphism $H^1(\Omega_k,\mu_n) \xrightarrow{\cong} \mathbb{Z}/n$. There results an embedding $H^1(\Omega,\mu_n) \hookrightarrow \prod_{\vec{k} \in P(\mathfrak{T})} (\mathbb{Z}/n)_{\vec{k}}$ with values in the group of alternating maps. Again by (A.4) we see that the image consists precisely of the set of harmonic elements. Clearly, the resulting isomorphism is compatible with the operation of $\Gamma$. □

For a detailed proof of (A.7), see [13, pp.175-182].

Next assume $\Gamma$ to be p'-torsion free. Hence $\Gamma$ acts on $\Omega$ without fixed points, and the stabilizers $\Gamma_i$ of vertices resp. edges of $\mathfrak{T}$ are p-groups. Applying (A.5) to all the pairs $(\Omega_i, \Gamma_i)$ and glueing the $\Gamma_i \backslash \Omega_i$ to $\Gamma \backslash \Omega$, we obtain an exact sequence

(A.8) $$0 \to H^1(\Gamma\backslash\mathfrak{T}, \mu_n) \to H^1(\Gamma\backslash\Omega, \mu_n) \to \underline{H}^1(\mathfrak{T}, \mathbf{Z}/n)^\Gamma \to 0 ,$$

where $(\ldots)^\Gamma$ denotes invariants of $\Gamma$, and $H^1(\Gamma\backslash\mathfrak{T}, \mu_n)$ is graph cohomology with values in the group $\mu_n$. All the groups occurring in (A.8) are finite by (2.3). Comparing with the topological analogue (1.10), (A.8) corresponds to the sequence $0 \to E_2^{1,0} \to E^1 \to E_2^{0,1} \to \ldots$ of terms of low degree of the Leray spectral sequence of $\lambda_\Gamma$.

Now consider $c \in \underline{H}^1(\mathfrak{T}, \mathbf{Z}/n)^\Gamma$ as an alternating function on $P(\Gamma\backslash\mathfrak{T})$. Let $\underline{H}^1_!(\mathfrak{T}, \mathbf{Z}/n)^\Gamma$ be the subgroup of those $c$ that vanish on the ends of $\Gamma\backslash\mathfrak{T}$, and let $H^1_!(\Gamma\backslash\Omega, \mu_n)$ be the pre-image. It consists of the invertible sheafs trivial on a neighborhood of the cusps; by (2.4) and (A.3), this agrees with the usual étale cohomology of algebraic curves

$$H^1(\bar{M}_\Gamma, \mu_n) \hookrightarrow H^1(M_\Gamma, \mu_n) = H^1(\Gamma\backslash\Omega, \mu_n) .$$

We observe:

(A.9) (i) $\Gamma$ being p'-torsion free and $(n,p) = 1$, we have

$$H^1(\Gamma\backslash\mathfrak{T}, \mathbf{Z}/n) = H^1(\Gamma, \mathbf{Z}/n) = \mathrm{Hom}(\Gamma, \mathbf{Z}/n) .$$

(ii) Putting $n = l^r$ with some prime number $l \neq p$, we have

$$\varprojlim_r H^1(\Gamma\backslash\mathfrak{T}, \mathbf{Z}/l^r) = H^1(\Gamma\backslash\mathfrak{T}, \mathbf{Z}) \otimes \mathbf{Z}_l ,$$

and the corresponding statement is true for $\underline{H}^1_!(\ldots)^\Gamma$.

Tensoring with $\mu_n^{-1}$ and going over to $\mathbb{Q}_l$-cohomology, we obtain from (A.8) the exact sequence

(A.10) $$0 \to H^1(\Gamma\backslash\mathfrak{T}, \mathbb{Q}) \otimes \mathbb{Q}_l \to H^1(\bar{M}_\Gamma, \mathbb{Q}_l) \to \underline{H}^1_!(\mathfrak{T}, \mathbb{Q})^\Gamma \otimes \mathbb{Q}_l(-1) \to 0 .$$

As is easy to verify, the dimensions of the exterior vector spaces agree: $\underline{H}^1_!(\mathfrak{T}, \mathbb{Q})^\Gamma$ is non-canonically isomorphic with $H^1(\Gamma\backslash\mathfrak{T}, \mathbb{Q})$, having the first Betti number of $\Gamma\backslash\mathfrak{T}$ as its dimension.

A.11 <u>Conclusion</u>. The genus $g(\bar{M}_\Gamma)$ of $\bar{M}_\Gamma$ agrees with the dimension $b_1(\Gamma)$ of $H^1(\Gamma,\mathbb{Q})$, and this fact remains true for arbitrary arithmetic subgroups $\Gamma'$ of $GL(2,K)$. For let $\Gamma \subset \Gamma'$ be a normal p'-torsion free subgroup of finite index. For $\Gamma$, we have (A.8) and (A.10). Choose $\mathfrak{n}$ resp. $l$ prime to $p$ and to the index $[\Gamma' : \Gamma]$. Considering the invariants of $\Gamma'/\Gamma$, we get (A.10) for $\Gamma'$. Clearly, this is true for the exterior terms of (A.10); for the middle term, we observe the assertion of (A.5) to hold true even for the ramified covering $\bar{M}_\Gamma \to \bar{M}_{\Gamma'}$, since the index $[\Gamma' : \Gamma]$ is prime to $l$.

A.12 <u>Remark</u>. The cohomology module $H_\Gamma = H^1(\bar{M}_\Gamma,\mathbb{Q}_l)$ has a much richer structure than used here. For reasons of simplicity, we only considered analytic spaces above $C$. In fact, the analytic structure of $\Omega$ and of all the manifolds considered is already defined over $K_\infty$. Therefore, one obtains a $\mathrm{Gal}(\bar{K}_\infty : K_\infty)$ - action on $H_\Gamma$. Regarding $\bar{M}_\Gamma$ as a component of the set of C-valued points of some modular scheme $\bar{M}(\mathfrak{n})$ and going over to the projective limit over all the divisors $\mathfrak{n}$, even the group $GL(2,\mathbb{A}_f) \times \mathrm{Gal}(\bar{K} : K)$ acts on the corresponding cohomology module. Proceeding, Drinfeld obtains a reciprocity law relating representations of $GL(2,\mathbb{A}_f)$ and two-dimensional l-adic representations of $\mathrm{Gal}(\bar{K} : K)$ [11, Thm.2].

## <u>VI Expansions around Cusps</u>

### <u>1. Preparations</u>

In the whole chapter excepted section 5, we use the following notations: $\mathfrak{a}$ and $\mathfrak{b}$ are ideals (oBdA assumed to be integral), $Y$ is the lattice $\mathfrak{a}(1,0) + \mathfrak{b}(0,1) \subset K^2$, $\Gamma = GL(Y)$, $\Lambda = Y_\omega = \mathfrak{a}\omega+\mathfrak{b}$ the corresponding lattice in $C$, depending on $\omega \in \Omega$, $\mathfrak{n} \subset A$ an integral ideal, $0 \neq f \in \mathfrak{n}$, and $\mathfrak{m} = (f)\mathfrak{n}^{-1}$.

Our first goal is to compute the expansion of $\Delta_\mathfrak{n}(\omega)$ around the cusp $\infty$ of $\bar{M}_\Gamma$. This will be done by a strong generalization of the methods used in [19,20]. First, we note:

$$\Gamma = \left\{ \begin{pmatrix} a & b \\ c & d \end{pmatrix} \in G(K) \;\middle|\; \begin{matrix} a,d \in A \;,\; ad-bc \in \mathbb{F}_q^* \\ b \in \mathfrak{a}^{-1}\mathfrak{b} \;,\; c \in \mathfrak{a}\mathfrak{b}^{-1} \end{matrix} \right\} .$$

By (V 2.6), $e^{1-q}_{\mathfrak{a}^{-1}\mathfrak{b}}$ is a parameter at $\infty$. We begin by computing the expansion of $\Delta_{\mathfrak{n}}$ with respect to $e^{-1}_{\mathfrak{a}^{-1}\mathfrak{b}}$, suitably normalized such that it has better rationality properties. This normalization corresponds to the change $\exp(x) \longmapsto \exp(2\pi i x)$ in the classical situation.

For a fractional ideal $\mathfrak{g}$, put

(1.1) $$t_{\mathfrak{g}} = \xi^{-1}(\mathfrak{g}) e_{\mathfrak{g}}^{-1}(\omega) = e_{(\mathfrak{g})}^{-1}(\xi(\mathfrak{g})\omega) ,$$

$e_{(\mathfrak{g})}$ denoting the exponential function of the distinguished 1-lattice $\xi(\mathfrak{g}) \cdot \mathfrak{g} = \Lambda^{(\mathfrak{g})}$ (IV 5.1). Let further

(1.2) $$R_{\mathfrak{n}}^{(\mathfrak{g})}(X) = \rho_{\mathfrak{n}}^{(\mathfrak{g})}(X^{-1}) X^{q^{\deg \mathfrak{n}}}$$

with our special 1-D-modules $\rho^{(\mathfrak{g})}$. Then:

(1.3) (i) $R_{\mathfrak{n}}^{(\mathfrak{g})}$ is a polynomial of degree $q^{\deg \mathfrak{n}}-1$ with coefficients in $\tilde{B}$;

(ii) $R_{\mathfrak{n}}^{(\mathfrak{g})}(0) = 1(\rho_{\mathfrak{n}}^{(\mathfrak{g})}) = 1$;

(iii) $R_{\mathfrak{n}}^{(\mathfrak{g})}$ is in fact a polynomial in $X^{q-1}$; the non-trivial exponents are of the form $q^{\deg \mathfrak{n}}-q^i$, where $0 \leq i \leq \deg \mathfrak{n}$.

1.4 <u>Lemma</u>. Let $\mathfrak{n}$ be of degree $d$, $0 \leq r \leq 1$ and $z \in C$ with $|z| \leq r \cdot q^{-Q}$, $Q = \delta q^{g+\delta-1}/(q^{\delta}-1)$. Then $|R_{\mathfrak{n}}^{(\mathfrak{g})}(z)-1| \leq r q^{d-1}(q-1)$.

<u>Proof</u>. $R_{\mathfrak{n}}^{(\mathfrak{g})}(X)$ is the product $\prod (1-yX)$, where $y$ runs through the set of zeroes of $\rho_{\mathfrak{n}}^{(\mathfrak{g})}$. The absolute values $|y|$ are bounded by $q^Q$ (IV 4.13). If $s_i\{y\}$ denotes the i-th symmetric function in the $\{y\}$, we have $|s_i\{zy\}| \leq |z|^i q^{Qi} \leq r^i$. Further, $|R_{\mathfrak{n}}^{(\mathfrak{g})}(z)-1| \leq \sup_{i\geq 1} |s_i\{zy\}|$. Since the $s_i$ vanish for $i < q^{d-1}(q-1)$, we are done. □

Consider the isogeny

$$\phi_{\mathfrak{n}}^{\Lambda} = \sum_{i \leq 2 \deg \mathfrak{n}} l_i \tau^i$$

of (V 3.4), where $l_{2 \deg \mathfrak{n}} = \Delta_{\mathfrak{n}}$. Regarding it as an additive polynomial (i.e. replacing $\tau^i$ by $X^{q^i}$), we obtain

$$\phi_{\mathfrak{n}}^{\Lambda}(X) = \Delta_{\mathfrak{n}} \prod (X-e_u) ,$$

$e_u$ as in (V 3.7), and $u$ running through a RS of $\mathfrak{n}^{-1}Y/Y$. This latter will be chosen in the form $u = (a/f, b/f)$, the $a$ (resp. the $b$) for their part running through a RS of $\mathfrak{m}\mathfrak{a} \bmod \mathfrak{f}\mathfrak{a}$ (resp. of $\mathfrak{m}\mathfrak{b} \bmod \mathfrak{f}\mathfrak{b}$) given by a $\mathbb{F}_q$-vector space section (see IV 3.1)). Then

$$(1.5) \qquad \Delta_{\mathfrak{n}}(\omega) \prod_u{}' e_u(\omega) = D(\phi_{\mathfrak{n}}^{\Lambda}) = 1 ,$$

which will be used to compute $\Delta_{\mathfrak{n}}$.

## 2. Formulae

In order not to interrupt the course of the argument, we collect here some calculations needed later. The following equations are to be understood in $C\{\tau\}$, note (IV 5.3). To limit the monstrosity of the formulae, we write $\rho z$ for the value $\rho(z)$ of the additive polynomial $\rho$ at $z$, provided this will cause no confusion.

As a direct consequence of the definition,

$$(2.1) \qquad t_{\mathfrak{g}}(c\omega) = t_{c^{-1}\mathfrak{g}}(\omega) \qquad (c \in K^*) .$$

Combining (ii) and (iii) of (IV 5.4), we get

$$(2.2) \qquad \rho_{\mathfrak{f}}^{(\mathfrak{b})} = \Theta(\mathfrak{m},\mathfrak{m}\mathfrak{b})\,\rho_{\mathfrak{m}}^{(\mathfrak{m}\mathfrak{b})}\,\Theta(\mathfrak{n},\mathfrak{b})\,\rho_{\mathfrak{n}}^{(\mathfrak{b})} .$$

Further,

$$(2.3) \qquad t_{\mathfrak{b}}^{-1}(f\omega) = \Theta(\mathfrak{m},\mathfrak{m}\mathfrak{b})\,\rho_{\mathfrak{m}}^{(\mathfrak{m}\mathfrak{b})}\,t_{\mathfrak{n}^{-1}\mathfrak{b}}^{-1}(\omega) .$$

<u>Proof.</u>

$$\begin{aligned} t_{\mathfrak{b}}^{-1}(f\omega) &= e_{(\mathfrak{b})}(\xi(\mathfrak{b})f\omega) \\ &= \rho_{\mathfrak{f}}^{(\mathfrak{b})} e_{(\mathfrak{b})}(\xi(\mathfrak{b})\omega) \\ &= \Theta(\mathfrak{m},\mathfrak{m}\mathfrak{b})\,\rho_{\mathfrak{m}}^{(\mathfrak{m}\mathfrak{b})}\,\Theta(\mathfrak{n},\mathfrak{b})\,\rho_{\mathfrak{n}}^{(\mathfrak{b})} e_{(\mathfrak{b})}(\xi(\mathfrak{b})\omega) \\ &= \Theta(\mathfrak{m},\mathfrak{m}\mathfrak{b})\,\rho_{\mathfrak{m}}^{(\mathfrak{m}\mathfrak{b})} e_{(\mathfrak{n}^{-1}\mathfrak{b})}(\Theta(\mathfrak{n},\mathfrak{b})D(\rho_{\mathfrak{n}}^{(\mathfrak{b})})\xi(\mathfrak{b})\omega) \qquad \text{(IV 5.5)} \\ &= \Theta(\mathfrak{m},\mathfrak{m}\mathfrak{b})\,\rho_{\mathfrak{m}}^{(\mathfrak{m}\mathfrak{b})}\,t_{\mathfrak{n}^{-1}\mathfrak{b}}^{-1}(\omega) \quad \text{by (IV 5.4i)} \qquad \square \end{aligned}$$

Putting $\omega/f$ for $\omega$ and $\mathfrak{b}\mathfrak{n}$ for $\mathfrak{b}$ in (2.3), we obtain

(2.4) $$t_{\mathfrak{b}\mathfrak{n}}^{-1}(\omega) = \Theta(\mathfrak{m},\mathfrak{b})\rho_{\mathfrak{m}}^{(\mathfrak{b})}t_{\mathfrak{b}}^{-1}(\omega/f) .$$

On the other hand,

(2.5) $$t_{\mathfrak{b}}^{-1}(\omega) = \Theta(\mathfrak{n},\mathfrak{n}\mathfrak{b})\rho_{\mathfrak{n}}^{(\mathfrak{n}\mathfrak{b})}t_{\mathfrak{n}\mathfrak{b}}^{-1}(\omega)$$

by substituting $\omega/f$ for $\omega$ in (2.3), applying (2.1) and interchanging the parts of $\mathfrak{n}$ and $\mathfrak{m}$ .

## 3. Computation of the Factors

Let $u = (u_1,u_2) = (a/f,b/f)$ , where not both $a$ and $b$ equal zero. We have

(3.1) $$e_u(\omega) = e_\Lambda(u_1\omega+u_2) = (u_1\omega+u_2) \prod_{\substack{c \in \mathfrak{a} \\ d \in \mathfrak{b}}}{}' \left(1-\frac{u_1\omega+u_2}{c\omega+d}\right) .$$

Let $c \neq 0$ . The function $z \longmapsto \prod_d (1-z/(c\omega+d))$ has simple zeroes for $z-c\omega \in \mathfrak{b}$ and no further zeroes, so has to be proportional to $e_{\mathfrak{b}}(z-c\omega)$ (Non-archimedean analysis!). Evaluating at $z = 0$ gives

$$\prod_d (1-z/(c\omega+d)) = \frac{e_{\mathfrak{b}}(z-c\omega)}{e_{\mathfrak{b}}(-c\omega)} ,$$

i.e. the factor corresponding to $c$ in (3.1) is

(3.2) $$\frac{e_{\mathfrak{b}}(u_1\omega+u_2-c\omega)}{e_{\mathfrak{b}}(-c\omega)} = \frac{e_{(\mathfrak{b})}(\xi(\mathfrak{b})(cf-a)\omega/f)-e_{(\mathfrak{b})}(\xi(\mathfrak{b})b/f)}{e_{(\mathfrak{b})}(\xi(\mathfrak{b})c\omega)} .$$

Here, we have multiplied the numerator and the denominator of the left hand side by $\xi(\mathfrak{b})$ . Note that $cf-a \in \mathfrak{m}\mathfrak{a}$ , and

$$z(b/f,\mathfrak{b}) = e_{(\mathfrak{b})}(\xi(\mathfrak{b})b/f)$$

is a $\mathfrak{n}$-division point of $\rho^{(\mathfrak{b})}$ .

Next, we have to express all the terms in (3.2) by $t_{\mathfrak{c}}(\omega)$, where $\mathfrak{c} = \mathfrak{a}^{-1}\mathfrak{b}\mathfrak{n}$. Let us first treat the denominator. We have

$$e_{(\mathfrak{b})}(\xi(\mathfrak{b})c\omega) = \Theta(\mathfrak{a}',\mathfrak{a}^{-1}\mathfrak{b})\rho_{\mathfrak{a}'}^{(\mathfrak{a}^{-1}\mathfrak{b})}e_{(\mathfrak{a}^{-1}\mathfrak{b})}(\xi(\mathfrak{a}^{-1}\mathfrak{b})\omega)$$

(where we have put $(c) = \mathfrak{a}\mathfrak{a}'$ and used (2.3))

$$= \Theta(\mathfrak{a}',\mathfrak{a}^{-1}\mathfrak{b})\rho_{\mathfrak{a}'}^{(\mathfrak{a}^{-1}\mathfrak{b})}\Theta(\mathfrak{n},\mathfrak{a}^{-1}\mathfrak{b}\mathfrak{n})\rho_{\mathfrak{n}}^{(\mathfrak{a}^{-1}\mathfrak{b}\mathfrak{n})}t_{\mathfrak{c}}^{-1}$$

$$= \Theta(\mathfrak{a}'\mathfrak{n},\mathfrak{c})\rho_{\mathfrak{a}'\mathfrak{n}}^{(\mathfrak{c})}t_{\mathfrak{c}}^{-1}$$

(3.3) $$= \Theta((c)\mathfrak{n}\mathfrak{a}^{-1},\mathfrak{c})\rho_{(c)\mathfrak{n}\mathfrak{a}^{-1}}^{(\mathfrak{c})}t_{\mathfrak{c}}^{-1} .$$

(To obtain the last three equalities, we applied (2.5) and (IV 5.4ii).) By $cf-a \in \mathfrak{m}\mathfrak{a}$, we may write

$$(cf-a) = \mathfrak{m}\mathfrak{a}\mathfrak{h}$$

with a certain positive divisor $\mathfrak{h}$. Then we get for the numerator term

$$e_{(\mathfrak{b})}(\xi(\mathfrak{b})(cf-a)\omega/f) = \rho_{cf-a}^{(\mathfrak{b})}e_{(\mathfrak{b})}(\xi(\mathfrak{b})\omega/f)$$

$$= \Theta(\mathfrak{a}\mathfrak{h},\mathfrak{b}\mathfrak{n})\rho_{\mathfrak{a}\mathfrak{h}}^{(\mathfrak{b}\mathfrak{n})}\Theta(\mathfrak{m},\mathfrak{b})\rho_{\mathfrak{m}}^{(\mathfrak{b})}e_{(\mathfrak{b})}(\xi(\mathfrak{b})\omega/f) \qquad (2.2)$$

$$= \Theta(\mathfrak{a}\mathfrak{h},\mathfrak{b}\mathfrak{n})\rho_{\mathfrak{a}\mathfrak{h}}^{(\mathfrak{b}\mathfrak{n})}t_{\mathfrak{b}\mathfrak{n}}^{-1} \qquad (2.4)$$

$$= \Theta(\mathfrak{h},\mathfrak{c})\rho_{\mathfrak{h}}^{(\mathfrak{c})}\Theta(\mathfrak{a},\mathfrak{b}\mathfrak{n})\rho_{\mathfrak{a}}^{(\mathfrak{b}\mathfrak{n})}t_{\mathfrak{b}\mathfrak{n}}^{-1} \qquad (\text{IV } 5.4\text{ii})$$

(3.4) $$= \Theta(\mathfrak{h},\mathfrak{c})\rho_{\mathfrak{h}}^{(\mathfrak{c})}t_{\mathfrak{c}}^{-1} . \qquad (2.5)$$

Substituting (3.3) and (3.4) into (3.2), the c-factor of (3.1) becomes

(3.5) $$\frac{\Theta(\mathfrak{h},\mathfrak{c})\rho_{\mathfrak{h}}^{(\mathfrak{c})}t_{\mathfrak{c}}^{-1} - z(b/f,\mathfrak{b})}{\Theta((c)\mathfrak{n}\mathfrak{a}^{-1},\mathfrak{c})\rho_{(c)\mathfrak{n}\mathfrak{a}^{-1}}^{(\mathfrak{c})}t_{\mathfrak{c}}^{-1}} ,$$

where $\mathfrak{h} = (cf-a)\mathfrak{m}^{-1}\mathfrak{a}^{-1}$ .

We multiply the numerator and denominator by $t_{\mathfrak{c}}$ to their degrees. Clearing factors, (3.5) becomes

$$(3.6)\qquad t_{\mathfrak{c}}^{-k}\quad \frac{\Theta(\mathfrak{h},\mathfrak{c})R_{\mathfrak{h}}^{(\mathfrak{c})}(t_{\mathfrak{c}}) - z(b/f,\mathfrak{b})t_{\mathfrak{c}}^{k'}}{\Theta((c)\mathfrak{n}a^{-1},\mathfrak{c})R_{(c)\mathfrak{n}a^{-1}}^{(\mathfrak{c})}(t_{\mathfrak{c}})}$$

with the polynomials $R_*^*$ of (1.2) and

$$k = k' - q^{\deg c + \deg \mathfrak{n} - \deg \mathfrak{a}}$$

$$k' = q^{\deg(cf-a)} - \deg \mathfrak{m} - \deg \mathfrak{a} .$$

3.7 <u>Proposition</u>. (i) The numerator of (3.6) has coefficients in $\tilde{B}(\mathfrak{n})$, whereas the coefficients of the denominator lie in $\tilde{B}$.

(ii) The exponent $k$ of $t_{\mathfrak{c}}$ vanishes for $\deg(cf) > \deg a$.

(iii) If $c$ satisfies $\deg(cf) > \deg a$, the $\Theta$-factors in the numerator and in the denominator agree. Thus, for such $c$, (3.6) is a power series in $t_{\mathfrak{c}}$ having 1 as its constant term.

<u>Proof</u>. (i) and (ii) are obvious. The equality of $\Theta$-factors up to $(q-1)$-st roots of unity follows from (IV 5.4iv). To get the identity, we go back to (3.6). Letting $t_{\mathfrak{h}}(\omega) \to 0$,

$$\frac{e_{(\mathfrak{h})}(\xi(\mathfrak{h})a\omega)}{e_{(\mathfrak{h})}(\xi(\mathfrak{h})cf\omega)} = \frac{\rho_a^{(\mathfrak{h})}t_{\mathfrak{h}}^{-1}}{\rho_{cf}^{(\mathfrak{h})}t_{\mathfrak{h}}^{-1}}$$

tends to zero as well, having a polynomial of smaller degree in the numerator. As one sees from (3.2), the whole expression tends to 1. □

It remains to treat the factor corresponding to $c = 0$ in (3.1). It is

$$(3.8)\qquad e_{\mathfrak{h}}(u_1\omega+u_2) = \xi(\mathfrak{h})^{-1}e_{(\mathfrak{h})}(\xi(\mathfrak{h})(a\omega+b)/f)$$

$$= \xi(\mathfrak{h})^{-1}[e_{(\mathfrak{h})}(\xi(\mathfrak{h})a\omega/f)+z(b/f,\mathfrak{b})] .$$

Proceeding as in (3.4), this becomes

$$\xi(\mathfrak{h})^{-1}[\Theta((a)\mathfrak{m}^{-1}\mathfrak{a}^{-1},\mathfrak{c})\rho_{(a)\mathfrak{m}^{-1}\mathfrak{a}^{-1}}^{(\mathfrak{c})}t_{\mathfrak{c}}^{-1} + z(b/f,\mathfrak{b})]$$

$$= \xi(\mathfrak{h})^{-1}t_{\mathfrak{c}}^{-k}[\Theta(\ldots)R_{(a)\mathfrak{m}^{-1}\mathfrak{a}^{-1}}^{(\mathfrak{c})}(t_{\mathfrak{c}}) + z(b/f,\mathfrak{b})t^k] ,$$

where in case $a \neq 0$, $k$ takes the value $q^{\deg a - \deg \mathfrak{m} - \deg \mathfrak{a}}$. If

$\mathfrak{a} = 0$ , the first term in [...] cancels, and we get $\xi(\mathfrak{b})^{-1} z(b/f,\mathfrak{b})$ . Putting for $c \neq 0$

$$\mathfrak{h}(c) = (c)\mathfrak{m}^{-1}\mathfrak{a}^{-1}$$

$$\Theta(c) = \Theta(\mathfrak{h}(c),\mathfrak{c}) ,$$

we have $(c)\mathfrak{n}\mathfrak{a}^{-1} = (f)\mathfrak{h}(c)$ , so

$$\Theta((c)\mathfrak{n}\mathfrak{a}^{-1},\mathfrak{c}) = \Theta((f)\mathfrak{h}(c),\mathfrak{c}) = \mathrm{sgn}(f,\mathfrak{b})\Theta(c) \qquad \text{(IV 5.4v)}$$

We put further $\Theta(0) = 0$ and $\deg \mathfrak{h}(0) = -\infty$ .

Combining (3.6) and (3.8) with (3.1) gives finally for $u = (u_1,u_2) = (a/f,b/f)$ :

$$\text{(3.9)} \qquad e_u(\omega) = \xi(\mathfrak{b})^{-1} t_{\mathfrak{c}}^{-k} \prod_{c \in \mathfrak{a}} F_c(t_{\mathfrak{c}})$$

with the factors

$$F_c = \Theta(a) R^{(\mathfrak{c})}_{\mathfrak{h}(a)}(t_{\mathfrak{c}}) + z(b/f,\mathfrak{b}) t_{\mathfrak{c}}^{q^{\deg \mathfrak{h}(a)}} \qquad (c = 0)$$

$$= \frac{\Theta(cf-a) R^{(\mathfrak{c})}_{\mathfrak{h}(cf-a)}(t_{\mathfrak{c}}) - z(b/f,\mathfrak{b}) t_{\mathfrak{c}}^{q^{\deg \mathfrak{h}(cf-a)}}}{\mathrm{sgn}(f,\mathfrak{b})\Theta(c) R^{(\mathfrak{c})}_{(f)\mathfrak{h}(c)}(t_{\mathfrak{c}})} \qquad (c \neq 0) ,$$

where

$$\mathfrak{c} = \mathfrak{a}^{-1}\mathfrak{b}\mathfrak{n} \quad \text{and}$$

$$k = q^{\deg \mathfrak{n} - \deg \mathfrak{a}}(Z_{u_1,\mathfrak{a}}(q) - Z_{0,\mathfrak{a}}(q)) .$$

The exponent $k$ of $t_{\mathfrak{c}}$ results from

$$k = \lim_{N\to\infty}[q^{\deg a - \deg \mathfrak{m} - \deg \mathfrak{a}} + \sum_{c \in \mathfrak{a}_N}{}' q^{\deg(cf-a) - \deg \mathfrak{m} - \deg \mathfrak{a}} - q^{\deg c + \deg \mathfrak{n} - \deg \mathfrak{a}}] ,$$

the identity

$$\{cf-a \mid 0 \neq c \in \mathfrak{a}_N\} = \{x \in \mathfrak{a}_{N+\deg f} \mid x \equiv -a \bmod f\,\mathfrak{a}, x \neq -a\} ,$$

valid for $N >> 0$ , and (III 3.11).

3.10 Remarks. (i) Let $Q$ be the constant of (1.4) and $0 < r < 1$. The product $\prod F_c(t_\mathfrak{c})$ converges for $|t_\mathfrak{c}| \le r \cdot q^{-Q}$. For, if $d = \deg(fc) - \deg a - \deg \mathfrak{m} > 0$, $F_c$ is a power series with constant term 1, and by (IV 4.13) and (1.4), the inequality $|F_c(t_\mathfrak{c}) - 1| \le r q^{d-1}(q-1)$ holds. Thus, (3.9) gives the Laurent expansion of the meromorphic function $e_u(\omega)$ in the pointed disc $0 < |t_\mathfrak{c}| \le r \cdot q^{-Q}$.

(ii) We always have $k \ge 0$, and $k = 0$ is equivalent with $u_1 = 0$. In particular, the Eisenstein series $E_u^{(1)} = e_u^{-1}$, as well as the coefficient functions $l_i$ of (V 3.4) are holomorphic at $t_\mathfrak{c} = 0$.

(iii) To each Laurent coefficient of $e_u$, only a finite number of the factors $F_c$ contribute. This follows from (1.3 iii). Consequently, the numbers $\tilde{a}_i = \xi(\mathfrak{h})$ times the i-th Laurent coefficient of (3.9) lie in $B(\mathfrak{n})$.

(iv) By construction of $R_*^*$ and the $\Theta$-factors, the Galois group $\operatorname{Gal}(\tilde{H}(\mathfrak{n}) : H(\mathfrak{n})) \xrightarrow{\cong} k^*$ acts on $a_i$ by $(\sigma, a_i) \longmapsto \sigma^i \cdot a_i$ $(\sigma \in k^*)$.

## 4. The Δ-Functions

Starting from (1.5), we have

$$(4.1) \qquad \Delta_\mathfrak{n}(\omega) = \prod_u{}' e_u^{-1}(\omega) = \prod_{c \in a} (\text{factors derived from } (3.2)-(3.8)).$$

Let first $c \neq 0$. The corresponding factor is

$$(4.2) \qquad \prod_{a,b}{}' \frac{e_{(\mathfrak{h})}(\xi(\mathfrak{h})c\omega)}{e_{(\mathfrak{h})}(\xi(\mathfrak{h})(cf-a)\omega/f) - z(b/f,\mathfrak{h})}.$$

We will express the numerator and denominator as polynomials in $t_\mathfrak{c}^{-1}$, where now $\mathfrak{c} = \mathfrak{a}^{-1}\mathfrak{h}$.

By the computation preceding (3.3), the numerator equals

$$(4.3) \qquad [\Theta((c)\mathfrak{a}^{-1}, \mathfrak{c})\, \rho^{(\mathfrak{c})}_{(c)\mathfrak{a}^{-1}} t_\mathfrak{c}^{-1}]^{q^{2\deg \mathfrak{n}}-1}.$$

Next, we treat the denominator of (4.2). For a fixed, take the product

$$(4.4) \qquad \prod_b (e_{(\mathfrak{b})}(\xi(\mathfrak{b})(cf-a)\omega/f) - z(b/f,\mathfrak{b}))$$

$$= \rho_{\mathfrak{n}}^{(\mathfrak{b})}(e_{(\mathfrak{b})}(\xi(\mathfrak{b})(cf-a)\omega/f)) \qquad \text{(II 3.7)}$$

$$= \rho_{\mathfrak{n}}^{(\mathfrak{b})} t_{\mathfrak{b}}^{-1}((cf-a)\omega/f) \ .$$

Writing again $(cf-a) = \mathfrak{a} \cdot \mathfrak{m} \cdot \mathfrak{h}(cf-a)$ , this becomes

$$= \Theta^{-1}(\mathfrak{n},\mathfrak{b}) t_{\mathfrak{b}\mathfrak{m}}^{-1}((cf-a)\omega) \qquad \text{((2.4) with the parts of } \mathfrak{n} \text{ and } \mathfrak{m} \text{ interchanged)}$$

$$= \Theta^{-1}(\mathfrak{n},\mathfrak{b})\Theta(\mathfrak{h},\mathfrak{c})\rho_{\mathfrak{h}}^{(\mathfrak{c})} t_{\mathfrak{c}}^{-1} \ .$$

For the last equation, we have used (2.3), the data $(\mathfrak{b},f,\mathfrak{n},\mathfrak{m})$ replaced by $(\mathfrak{b}\mathfrak{m},cf-a,\mathfrak{a}\mathfrak{m},\mathfrak{h})$ .

Putting $t = t_{\mathfrak{c}}$ , where $\mathfrak{c} = \mathfrak{a}^{-1}\mathfrak{h}$ , and $\Theta(x) = \Theta((x)\mathfrak{a}^{-1}\mathfrak{m}^{-1},\mathfrak{c})$ , the factor corresponding to $c \neq 0$ in (4.1) is given by

$$(4.5) \qquad \frac{[\Theta((c)\mathfrak{a}^{-1},\mathfrak{c})\rho_{(c)\mathfrak{a}^{-1}}^{(\mathfrak{c})} t^{-1}]^{q^{2 \deg \mathfrak{n}}}}{\prod_a \Theta^{-1}(\mathfrak{n},\mathfrak{b})\Theta(cf-a)\rho_{\mathfrak{h}(cf-a)}^{(\mathfrak{c})} t^{-1}}$$

$$= t^k \ \frac{[\Theta((c)\mathfrak{a}^{-1},\mathfrak{c})R_{(c)\mathfrak{a}^{-1}}^{(\mathfrak{c})}(t)]^{q^{2 \deg \mathfrak{n}}}}{\prod_a \Theta^{-1}(\mathfrak{n},\mathfrak{b})\Theta(cf-a)R_{\mathfrak{h}(cf-a)}^{(\mathfrak{c})}(t)} \ ,$$

where $k = \sum_a q^{\deg(cf-a) - \deg \mathfrak{a} - \deg \mathfrak{m}} - q^{2 \deg \mathfrak{n} + \deg c - \deg \mathfrak{a}}$ .

According to (3.7), we have:

(4.6) (i) The coefficients of (4.5) lie in $\tilde{B}$ ;

(ii) The exponent $k$ vanishes for $\deg c >> 0$ , and for such $c$ , (4.5) is a power series with constant term 1.

The factor corresponding to $c = 0$ in (4.1) is given by

$$(4.7) \qquad \prod_{a,b}{}' e_{\mathfrak{b}}^{-1}(u_1\omega+u_2) = \prod{}' \xi(\mathfrak{b}) e_{(\mathfrak{b})}^{-1}(\xi(\mathfrak{b})(a\omega+b)/f) \ .$$

For $a \neq 0$ ,

$$\prod_b e_{(\mathfrak{h})}(\xi(\mathfrak{h})(a\omega+b)/f) = \rho_{\mathfrak{n}}^{(\mathfrak{h})} e_{(\mathfrak{h})}(\xi(\mathfrak{h})a\omega/f)$$

$$= \Theta^{-1}(\mathfrak{n},\mathfrak{h})\Theta(a)\rho^{(\mathfrak{c})}_{\mathfrak{h}(a)} t^{-1} ,$$

and finally,

$$\prod_b{}' e_{(\mathfrak{h})}(\xi(\mathfrak{h})b/f) = \prod_b{}' z(b/f,\mathfrak{h})$$

$$= D(\rho_{\mathfrak{n}}^{(\mathfrak{h})}) .$$

Thus, we obtain for (4.7) the following expression

$$(4.8) \qquad \xi(\mathfrak{h})^{q^{2\deg \mathfrak{n}}-1} D^{-1}(\rho_{\mathfrak{n}}^{(\mathfrak{h})})\Theta(\mathfrak{n},\mathfrak{h})^{q^{\deg \mathfrak{n}}-1}[\prod_a{}'\Theta(a)\rho^{(\mathfrak{c})}_{\mathfrak{h}(a)} t^{-1}]^{-1}$$

$$= \xi(b)^{q^{2\deg \mathfrak{n}}-1} D^{-1}(\rho_{\mathfrak{n}}^{(\mathfrak{h})})\Theta(\mathfrak{n},\mathfrak{h})^{q^{\deg \mathfrak{n}}-1} t^k [\prod_a{}'\Theta(a)R^{(\mathfrak{c})}_{\mathfrak{h}(a)}(t)]^{-1} ,$$

where $k = \sum_a{}' q^{\deg a - \deg \mathfrak{m} - \deg \mathfrak{a}}$ .

We may now take the product of the contributions of $c \in \mathfrak{a}$, the convergence resulting from that of (3.9). If $c$ runs through $\mathfrak{a} - \{0\}$ and $a$ through our RS of $\mathfrak{ma}$ mod $f\mathfrak{a}$, the elements $cf-a$ run through all the elements of $\mathfrak{ma} - \{\mathfrak{a}\}$. The exponent $k$ of $t$ in the product over all $c$ is therefore given by

$$(4.9) \qquad k = \lim_{N\to\infty} \sum_{c\in \mathfrak{a}_N}{}' \left( \sum_a q^{\deg(cf-a)-\deg a-\deg \mathfrak{m}} - q^{2\deg \mathfrak{n} + \deg c - \deg \mathfrak{a}} \right)$$

$$+ \sum_a{}' q^{\deg a - \deg \mathfrak{m} - \deg \mathfrak{a}}$$

$$= (q-1)(Z_{(\mathfrak{a}^{-1}\mathfrak{n})}(q) - q^{2\deg \mathfrak{n}} Z_{(\mathfrak{a}^{-1})}(q))$$

(see (III 3.10)).

Let now $R_{\mathfrak{h}} = R_{\mathfrak{h}}^{(\mathfrak{c})}$ . The R-parts of the numerator resp. denominator in the product of the contributions (4.5) resp. (4.8) are converging separately, so that we may write for the R-part:

$$(4.10) \qquad \prod_{c\in\mathfrak{a}}{}' R^{q^{2\deg \mathfrak{n}}}_{(c)\mathfrak{a}^{-1}}(t) \Big/ \prod_{c\in \mathfrak{am}}{}' R_{\mathfrak{h}(c)}(t)$$

$$= \prod_{\substack{\mathfrak{g} > 1 \\ \mathfrak{g} \sim \mathfrak{a}^{-1}}} R_{\mathfrak{g}}^{q^{2 \deg \mathfrak{n}}(q-1)}(t) \Big/ \prod_{\substack{\mathfrak{h} > 1 \\ \mathfrak{h} \sim \mathfrak{a}^{-1}\mathfrak{n}}} R_{\mathfrak{h}}^{q-1}(t) ,$$

each positive divisor $\mathfrak{g}$ equivalent with $\mathfrak{a}^{-1}$ occurring precisely $(q-1)$ times in the form $\mathfrak{g} = (c)\mathfrak{a}^{-1}$ (correspondingly for the $\mathfrak{h}$ ).

Taken together, we arrive at the product formula

(4.11) $\Delta_{\mathfrak{n}}(\omega) =$

$$\Theta D(\rho_{\mathfrak{n}}^{(\mathfrak{b})})^{-1} \xi(\mathfrak{b})^{q^{2 \deg \mathfrak{n}}-1} t^k \prod_{\mathfrak{g} \sim \mathfrak{a}^{-1}} R_{\mathfrak{g}}^{q^{2 \deg \mathfrak{n}}(q-1)}(t) \Big/ \prod_{\mathfrak{h} \sim \mathfrak{a}^{-1}\mathfrak{n}} R_{\mathfrak{h}}^{q-1}(t)$$

with some $(q^{\delta}-1)$-st root of unity $\Theta$ , $t = t_{\mathfrak{a}^{-1}\mathfrak{b}}$ , $R_* = R_*^{(\mathfrak{a}^{-1}\mathfrak{b})}$ , and $k$ is given by (4.9).

The product converges uniformly on discs $|t| \leq r \cdot q^{-Q}$ , where $0 < r < 1$ , and $Q$ is the constant occurring in (1.4).

By (V 2.6), the correct parameter at $\infty$ is $t^{q-1}$ , so $\Delta_{\mathfrak{n}}$ has a zero of order $Z_{(\mathfrak{a}^{-1}\mathfrak{n})}(q) - q^{2 \deg \mathfrak{n}} Z_{(\mathfrak{a}^{-1})}(q)$ at $\infty$ .

The root-of-unity factor $\Theta$ may be determined up to $(q-1)$-st roots of unity by (IV 5.4). I refrain from writing down the complicated formula. However, we have the following important special case, letting $\mathfrak{n} = (f)$ of degree $d$ . Then

(4.12) $$\Delta_f(\omega) = \Theta \xi(\mathfrak{b})^{q^{2d}-1} t^k \prod_{\mathfrak{g} \sim \mathfrak{a}^{-1}} R_{\mathfrak{g}}^{(q^{2d}-1)(q-1)}(t) ,$$

where

$$k = (1-q^{2d})(q-1) Z_{(\mathfrak{a}^{-1})}(q) \quad \text{and}$$

$$\Theta = \tau^{2-2g-\deg \mathfrak{a} - \deg \mathfrak{b}}(\mathrm{sgn}(f)) \cdot \tau^{1-g-\deg \mathfrak{b}}(\varepsilon_{(\mathfrak{a})}^{-d/\delta}) .$$

In this case, the factors $R_{\mathfrak{h}}$ in the denominator cancel, and $D(\rho)$ vanishes because of $\Delta_f = f \cdot \Delta_{(f)}$ and $f = \mathrm{sgn}(f,\mathfrak{b}) D(\rho_{(f)}^{(\mathfrak{b})})$ . The numbers $q^{2d}-1$ and $k$ being divisible by $q^{\delta}-1$ , the factors $\xi(\mathfrak{b})^{q^{2d}-1}$ , $\Theta$ , and $t^k$ do not depend on the choices (IV 5.1) of our distinguished 1-D-modules.

The value given for $\Theta$ (which is valid <u>precisely</u>, i.e. not only modulo $\mu_{q-1}$ !) is obtained by the following consideration: The formulae (3.2) and (3.8) show that $\xi(\mathfrak{h})^{-1}e_u^{-1}$ may be expanded as a power series in $t_{f\mathfrak{h}}$ such that the leading coefficient is a precisely computable number. Hence, the leading coefficient $b_j$ of

$$(4.13) \qquad \xi(\mathfrak{h})^{1-q^{2d}}\Delta_f = \sum b_i\, t_{f\mathfrak{h}}^i$$

is precisely computable (which will be carried out in Lemma 4.14). From (2.5), we get

$$t_{\mathfrak{a}^{-1}\mathfrak{h}} = t_{f\mathfrak{h}}^{q^{d+\deg\mathfrak{a}}} \,/\, \Theta((f)\mathfrak{a},\mathfrak{h})\, R^{(\mathfrak{h})}_{(f)\mathfrak{a}}(t_{f\mathfrak{h}}) \,.$$

Using $R_*^*(0) = 1$ and $k \equiv 0(q^\delta-1)$ , we see the equality $a_k = b_j$ (in fact, $j = k \cdot q^{d+\deg\mathfrak{a}}$ ) of leading coefficients of

$$\xi(\mathfrak{h})^{1-q^{2d}}\Delta_f = \sum a_i t^i$$

and of the expansion (4.13). Combined with (4.14), this will show the assertion on $\Theta$ .

4.14 <u>Lemma</u>. The leading coefficient $b_j$ in (4.13) is given by

$$b_j = \tau^{2-2g-\deg\mathfrak{a}-\deg\mathfrak{h}}(\operatorname{sgn}(f)) \cdot \tau^{1-g-\deg\mathfrak{h}}(\varepsilon_{(\mathfrak{a})}^{-d/\delta}) \,.$$

<u>Proof</u>. OBdA, we assume $\mathfrak{a}$ to be strictly contained in $A$ . Then $cf - a \in \mathbb{F}_q$ implies $c = a = 0$ (we keep the notations of section 3). Let $w_{c,u}$ be the leading coefficient of the c-factor (3.2) in case $c \neq 0$ (resp. of (3.8) if $c = 0$ ) , considered as a power series in $t_{f\mathfrak{h}}$ . Let $z = z(\mathfrak{b}/f,\mathfrak{h})$ . We write (3.2) in the form

$$\frac{e_{(\mathfrak{h})}(\xi(\mathfrak{h})(cf-a)\omega/f)}{e_{(\mathfrak{h})}(\xi(\mathfrak{h})c\omega)} = \frac{t_{f\mathfrak{h}}^{-1}((cf-a)\omega) - z}{t_{f\mathfrak{h}}^{-1}(cf\omega)}$$

$$= \frac{\rho^{(\mathfrak{h})}_{cf-a}(t_{f\mathfrak{h}}^{-1}) - z}{\rho^{(\mathfrak{h})}_{cf}(t_{f\mathfrak{h}}^{-1})}$$

$$= t_{f\mathfrak{h}}^{*}\,\tau^{1-g-\deg\mathfrak{h}}\left(\frac{\operatorname{sgn}(cf-a)}{\operatorname{sgn}(cf)}\right)(1+\ldots) \,,$$

i.e. $w_{c,u} = \tau^{1-g-\deg\mathfrak{h}}\left(\frac{\operatorname{sgn}(cf-a)}{\operatorname{sgn}(cf)}\right)$ , provided $c \neq 0$ .

Similarly, from

$$\begin{aligned} e_{(\mathfrak{b})}(\xi(\mathfrak{b})(a\omega+b)/f &= t_{\mathfrak{b}}^{-1}((a\omega+b)/f) \\ &= t_{f\mathfrak{b}}^{-1}(a\omega+b) \\ &= \rho_a^{(\mathfrak{b})}(t_{f\mathfrak{b}}^{-1}) + z \\ &= t_{f\mathfrak{b}}^{*}(\mathrm{sgn}(a,\mathfrak{b})R_{(a)}^{(\mathfrak{b})}(t_{f\mathfrak{b}}) + z\, t_{f\mathfrak{b}}^{q^{\deg a}}) , \end{aligned}$$

we get

$$\begin{aligned} w_{o,u} &= \xi(\mathfrak{b})^{-1}\mathrm{sgn}(a,\mathfrak{b}) \qquad (a \neq 0) \\ &\quad \xi(\mathfrak{b})^{-1}z(b/f,\mathfrak{b}) \qquad (a = 0) . \end{aligned}$$

The number $b_j$ to be determined is $\xi(\mathfrak{b})^{1-q^{2d}} f \cdot \lim_{N\to\infty} (\prod_u{}' \prod_{c \in \mathfrak{a}_N} w_{c,u}^{-1})$ .

Thus, let $N >> 0$ , and for abbreviation, $s(x) = \mathrm{sgn}(x,\mathfrak{b})$ . Then $\prod_u{}' \prod_{c \in \mathfrak{a}_N} w_{c,u}$

$$\begin{aligned} &= \prod_a{}'[\prod_b(\xi(\mathfrak{b})^{-1}s(a) \prod_{c \in \mathfrak{a}_N}{}' \frac{s(cf-a)}{s(cf)})] \cdot \prod_b{}' \xi(\mathfrak{b})^{-1}z(b/f,\mathfrak{b}) \\ &= \xi(\mathfrak{b})^{1-q^{2d}} D(\rho_{(f)}^{(\mathfrak{b})})\prod_a{}'[s(a)\prod_c{}' \frac{s(cf-a)}{s(cf)}] \qquad (*) . \end{aligned}$$

(The product of $s(a)$ over $b$ equals $s(a)$ , $d$ being divisible by $\delta$ .) We abbreviate further

$$D = \xi(\mathfrak{b})^{1-q^{2d}} D(\rho_{(f)}^{(\mathfrak{b})})$$

and obtain

$$\begin{aligned} (*) &= D(\prod_a{}' s(a)) \cdot \prod_a \prod_c{}' \frac{s(cf-a)}{s(cf)} \\ &= D \prod_{c \in \mathfrak{a}_{N+d}}{}' s(c) \Big/ \prod_{c \in \mathfrak{a}_N}{}' s(cf) \end{aligned}$$

(as above, $\prod_a s(cf) = s(cf)^{q^d} = s(cf)$ !)

$$= D\, \tau^{1-g-\deg \mathfrak{b}} (\varepsilon_{(a)}^{d/\delta}/s(f))^{\#(\mathfrak{a}_N)-1} .$$

For the last equation, we substituted the definition (IV 3.3) of $\varepsilon_{(\mathfrak{a})}$ . Taking into account

$$\dim \mathfrak{a}_N = 1-g-\deg \mathfrak{a} + N \equiv 1-g-\deg \mathfrak{a}\ (\delta) \quad \text{and}$$

$$f = s(f)D(\rho_{(f)}^{(\mathfrak{b})}) \ ,$$

$b_j$ comes out as asserted. □

4.15 Remark. We have the properties of $\Delta_{\mathfrak{n}}$ at $\infty$ corresponding to those described in (3.10):

(i) To each coefficient of the series expansion of (4.11), only a finite number of factors contribute;

(ii) The coefficients $a_i$ of $\xi(\mathfrak{b})^{1-q^{2\deg \mathfrak{n}}} \Delta_{\mathfrak{n}}(\omega)$ w.r.t. $t_{\mathfrak{a}^{-1}\mathfrak{b}}$ lie in $\tilde{B}$ ;

(iii) The Galois group $\mathrm{Gal}(\tilde{H} : H) \xrightarrow{\cong} k^*/\mathbb{F}_q^*$ operates on the $a_i$ by $(\sigma, a_i) \longmapsto \sigma^i \cdot a_i$ . $(\sigma \in k^*$ ; the numbers $a_i \neq 0$ satisfy $i \equiv 0 (q-1)!)$ .

## 5. Some Consequences

Let now $M(1) = M^2(1)$ be the coarse modular scheme for D-modules of rank 2. The components of $M(1)(C)$ resp. $\bar{M}(1)(C)$ are in bijective correspondence with Pic A , the cusps of $\bar{M}(1)(C)$ with Pic A × Pic A . We briefly describe this correspondence. Let G be the group scheme GL(2) and $\{\underline{x}\}$ a RS of the double coset $G(K)\backslash G(\mathbb{A}_f)/G(\hat{A})$ . By (V 2.4), we may extend the bijections (II 1.7, 1.8) to the diagram

(5.1)
$$\begin{array}{ccc} \bar{M}(1)(C) & \xrightarrow{\cong} & G(K)\backslash G(\mathbb{A}_f) \times (\Omega \cup \mathbb{P}_1(K))/G(\hat{A}) \\ & \searrow \quad \swarrow & \\ & \coprod_{\{\underline{x}\}} \Gamma_{\underline{x}} \backslash (\Omega \cup \mathbb{P}_1(K)) \ . & \end{array}$$

If $\underline{y} \in G(\mathbb{A}_f)$ can be written $\underline{y} = \gamma\, \underline{x}\, \underline{k}$ with some $\gamma \in G(K)$ and $\underline{k} \in G(\hat{A})$ , the right hand arrow assigns to the pair $(\underline{y}, z)$ the class of $\gamma^{-1}z$ in $\Gamma_x \backslash (\Omega \cup \mathbb{P}_1(K))^{\wedge}$. The cusps corresponding to the double coset $G(K)\backslash G(\mathbb{A}_f) \times \mathbb{P}_1(K)/G(\hat{A})$, we associate with the class of $(\underline{y},s)$ the element

$$(\text{class of } U(Y(\underline{y}),s) \text{ , class of } V(Y(\underline{y}),s)) \text{ of } \operatorname{Pic} A \times \operatorname{Pic} A .$$

This mapping is bijective by (II 1.4) and (V 2.4). Let $s(\mathfrak{a},\mathfrak{b})$ be the cusp corresponding to $((\mathfrak{a}),(\mathfrak{b})) \in \operatorname{Pic} A \times \operatorname{Pic} A$. Letting $Y = \mathfrak{a}(1,0) + \mathfrak{b}(0,1)$ and $\Gamma = GL(Y)$ as in sections 1 to 4, $s(\mathfrak{a},\mathfrak{b})$ lies on the component $\bar{M}_\Gamma$ and agrees with the cusp $\infty$.

In order to obtain a function on $\Omega$ from the function $\Delta_\mathfrak{n}$ a priori defined on the set of lattices, we have to fix a 2-lattice $Y \subset K^2$. (Up to now, $Y = \mathfrak{a}(1,0) + \mathfrak{b}(0,1)$ had been fixed, so we could omit $Y$ in the notation.)

Thus let

$$\Delta_\mathfrak{n}^Y(\omega) = 1(\phi_\mathfrak{n}^{Y_\omega}) .$$

For $\begin{pmatrix} a & b \\ c & d \end{pmatrix} = \nu \in G(K)$ and $k = q^{2 \deg \mathfrak{n}} - 1$,

(5.2) $$(c\omega+d)Y_{\nu\omega} = (Y\nu)_\omega \text{ , so}$$

(5.3) $$\begin{aligned} \Delta_\mathfrak{n}^Y[\nu]_k(\omega) &= (c\omega+d)^{-k} 1(\phi_\mathfrak{n}^{Y_{\nu\omega}}) \\ &= 1(\phi_\mathfrak{n}^{(Y\nu)_\omega}) \\ &= \Delta_\mathfrak{n}^{Y\nu}(\omega) . \end{aligned}$$

$Y$ being fixed, we may always find some $\nu$ satisfying $Y\nu = \mathfrak{a}(1,0) + \mathfrak{b}(0,1)$. Thus, by means of section 4, we are able to describe the modular forms $\Delta_\mathfrak{n}$ at all the cusps. In particular, we have shown:

5.4 Theorem. The modular form $\Delta_\mathfrak{n}$ has a zero of order $Z_{(\mathfrak{a}^{-1}\mathfrak{n})}(q) - q^{2 \deg \mathfrak{n}} Z_{(\mathfrak{a}^{-1})}(q)$ at the cusp $s(\mathfrak{a},\mathfrak{b})$.

Corresponding assertions hold for the functions $e_u$. For the rest of this section, we fix a lattice $Y \subset K^2$ and $\Gamma = GL(Y)$. Let $u \in \mathfrak{n}^{-1}Y - Y$. Then $e_u^Y$ is a modular form of weight $-1$ for $\Gamma(\mathfrak{n})$, and we have

(5.5) $$e^{Y}_{u[\nu]_{-1}}(\omega) = e^{Y\nu}_{u\nu}(\omega)$$

which is similar to (5.3).

Considering a cusp $s = \nu(\infty)$ of $\bar{M}_{\Gamma(\mathfrak{n})}$, we may assume $Y\nu = \mathfrak{a}'(1,0) + \mathfrak{b}'(0,1)$ with some ideals $\mathfrak{a}',\mathfrak{b}'$ of $A$, and (3.9) implies

5.6 Theorem. The meromorphic modular form $e^{Y}_{u}$ of $\Gamma(\mathfrak{n})$ has a pole of order $q^{\deg \mathfrak{n} - \deg \mathfrak{a}'}(Z_{u'_1,\mathfrak{a}'}(q) - Z_{0,\mathfrak{a}'}(q))$ at $s$, where $u' = (u'_1, u'_2) = u\nu$.

If $d = \deg \mathfrak{n}$, $\Delta_{\mathfrak{n}}$ is of weight $k = q^{2d}-1$. The line bundle $\mathfrak{M}_k$ of holomorphic modular forms of weight $k$ over $\bar{M}_{\Gamma}$ has the degree

(5.7) $$\begin{aligned} \deg(\mathfrak{M}_k) &= \sum_{(\mathfrak{a}) \in \operatorname{Pic} A} Z_{(\mathfrak{a}^{-1}\mathfrak{n})}(q) - q^{2d} Z_{(\mathfrak{a}^{-1})}(q) \\ &= (1-q^{2d}) Z_A(q) \\ &= \frac{(q^{2d}-1)(q^{\delta}-1)}{(q-1)(q^2-1)} P(q) . \end{aligned}$$

Using (V 5.5), we get

5.8 Theorem. The genus of the modular curve $\bar{M}_{\Gamma}$ is given by

$$g(\bar{M}_{\Gamma}) = 1 + (q^2-1)^{-1}\left[\frac{q^{\delta}-1}{q-1} P(q) - \frac{q(q+1)}{2}\,\delta \cdot P(1) + \eta\right] ,$$

where $\eta = 0$ for $\delta$ even and $\eta = -\frac{q(q-1)}{2} P(-1)$ for $\delta$ odd.

By (V A.11), $g(\bar{M}_{\Gamma}) = b_1(\Gamma) = \dim_{\mathbb{Q}} H^1(\Gamma,\mathbb{Q})$ . Some values of these numbers are given by the table

(5.9)

| $g$ = genus of $K$ | $\delta$ | $g(\bar{M}_\Gamma)$ |
|---|---|---|
| 0 | 1,2,3 | 0 |
| | 4 | $q$ |
| | 5 | $q(q+1)$ |
| | 6 | $q(q^2+q+2)$ |
| 1 | 1 | 0 |
| | 2 | $q^2$ |
| | 3 | $q(q^2+q-t)$ |

In the last example, we have put $P(X) = qX^2 - tX + 1$ .

By standard estimations of $P(q)$ and $P(1)$ , the degree $\frac{q^\delta - 1}{q-1} P(q)$ of the line bundle $\mathfrak{M}_{q^2-1}$ always exceeds $2g(\bar{M}_\Gamma)$ . This proves

5.10 Corollary. The line bundle $\mathfrak{M}_{q^2-1}$ on $\bar{M}_\Gamma$ is very ample and induces an imbedding $\bar{M}_\Gamma \hookrightarrow \mathbb{P}^N$ , where

$$N = \frac{q^\delta - 1}{q-1} P(q) - g(\bar{M}_\Gamma) .$$

In particular, we may compute $\dim M_k(\Gamma)$ by Riemann-Roch, provided that $k$ satisfies $k \equiv 0(q^2-1)$ .

Let $(\mathfrak{d})$ be the class of $\Lambda^2(Y)$ . We will identify

$$\operatorname{Pic} A \xrightarrow{\cong} \operatorname{Sp}(\Gamma)$$

$$\mathfrak{b} \longmapsto s(\mathfrak{d}\mathfrak{b}^{-1}, \mathfrak{b}) .$$

This in turn identifies the group ring $\mathbb{Z}[\operatorname{Pic} A]$ with the group of divisors on $\bar{M}_\Gamma$ supported by the cusps.

5.11 Theorem. Let $\{\mathfrak{n}\}$ be a RS of Pic A consisting of ideals $\mathfrak{n} \subsetneq A$ . The divisors of the $\Delta_\mathfrak{n}$ generate a subgroup of finite index in $\mathbb{Z}[\operatorname{Pic} A]$ .

Proof. We have

$$\operatorname{div}(\Delta_\mathfrak{n}) = \sum_{(\mathfrak{b}) \in \operatorname{Pic} A} (Z_{(\mathfrak{d}^{-1}\mathfrak{b}\mathfrak{n})}(q) - q^{2 \deg \mathfrak{n}} Z_{(\mathfrak{d}^{-1}\mathfrak{b})}(q))(\mathfrak{b}) ,$$

so we have to show the nonsingularity of the matrix

$$R((\mathfrak{b}),(\mathfrak{n})) = Z_{(\mathfrak{b}\mathfrak{n})}(q) - q^{2 \deg \mathfrak{n}} Z_{(\mathfrak{b})}(q) .$$

Now

$$R((\mathfrak{b}),(\mathfrak{n})) = S((\mathfrak{b}),(\mathfrak{n})) - q^{2 \deg \mathfrak{n}} S((\mathfrak{b}),(1)) , \text{ where}$$

$$S((\mathfrak{b}),(\mathfrak{n})) = Z_{(\mathfrak{b}\mathfrak{n})}(q) = \zeta_{(\mathfrak{b}\mathfrak{n})}(-1) .$$

Thus, it will suffice to show the nonsingularity of $S((\mathfrak{b}),(\mathfrak{n}))$ which, by the Frobenius determinant formula (e.g. [47, p.284]) is equivalent with the non-vanishing of all the L-values $L(\chi,-1)$ , $\chi$ running through the set of characters of Pic A . But, as in the number field case,

$$\prod L(\chi,-1) = \zeta_B(-1) = \zeta_H(-1)(1-q^{\delta})^h ,$$

where $\zeta_H$ is the Zeta function of H which does not vanish at $-1$ [68, p.130]. □

5.12 <u>Corollary</u>. The cusps of $\bar{M}_\Gamma$ generate a finite subgroup of the Jacobian.

<u>Proof</u>. The divisors of modular functions of the form $\prod \Delta_{\mathfrak{n}}^{i(\mathfrak{n})}$ generate a subgroup of finite index in the kernel of the degree mapping $\mathbb{Z}[\text{Pic } A] \to \mathbb{Z}$ . □

5.13 <u>Remark</u>. The same assertion may be proved for the cusps of congruence subgroups $\Gamma' \subset \Gamma$ . One has to consider those principal divisors which come from products of weight 0 of the forms $\Delta_{\mathfrak{n}}$ and $e_u$ . In the case of a rational function field, this has been carried out in [20]. Again, the proof depends on the non-vanishing of L-series at $s = -1$ , but a certain problem arises from the occurrence of "trivial" zeroes [20, 5.6], see also [45, Ch.I].

With the help of (4.12), we are able to construct a canonical modular form $\Delta$ of weight $j = q^{2\delta}-1$ . Let $f,f'$ be elements of A of degrees $d,d'$ , and let $i = q^{2d}-1$ , $i' = q^{2d'}-1$ . By construction,

$$\Delta_f \Delta_{f'}^{i+1} = \Delta_f^{i'+1} \Delta_{f'} , \text{ so}$$

$$\Delta_{f'}^{i} = \Delta_{f}^{i'} .$$

Choosing $d,d'$ with $(d,d') = \delta$ , we will have $(i,i') = j$ , see (IV 4.1). Now we write $j = ni+n'i'$ , and put

$$\Delta = \Delta_f^n \Delta_{f'}^{n'} .$$

Then $\Delta^i = \Delta_f^j$ , i.e. $\Delta_f$ and $\Delta^{i/j}$ agree up to $j$-th roots of unity which can be determined by (4.12). Therefore, we define

$$(5.14) \qquad \Delta = \Delta_f^n \Delta_{f'}^{n'} ,$$

choosing monic elements $f$ and $f'$ . The next proposition shows $\Delta$ to be well defined.

5.15 Proposition. (i) $\Delta$ does not depend on the choices of $f,f',n,n'$ . Using the notations of (4.11), it has the product expansion

$$\Delta = \Theta\xi(\mathfrak{b})^{q^{2\delta}-1} t^k \prod_{\mathfrak{g} \sim \mathfrak{a}^{-1}} R_{\mathfrak{g}}^{(q^{2\delta}-1)(q-1)}(t) ,$$

where

$$k = (1-q^{2\delta})(q-1)Z_{(\mathfrak{a}^{-1})}(q) \quad \text{and} \quad \Theta = \tau^{1-g-\deg \mathfrak{b}}(\varepsilon_{(\mathfrak{a})}^{-1}) .$$

(ii) $\Delta$ is a holomorphic modular form of weight $j = q^{2\delta}-1$ in the sense of (V 3.8).

(iii) $\Lambda \subset C$ denoting a 2-lattice whose second exterior power is isomorphic with $\mathfrak{d} \subset A$ , we have for each $f \in A$ of degree $d$

$$\Delta_f(\Lambda) = \tau^{2-2g-\deg \mathfrak{d}}(\mathrm{sgn}(f)) \cdot \Delta^{(q^{2d}-1)/j}(\Lambda) .$$

Proof. If $\varepsilon$ is a $(q^\delta-1)$-st root of unity, $\varepsilon^{d/\delta} = \varepsilon^{(q^d-1)/(q^\delta-1)} = \varepsilon^{(q^{2d}-1)/j}$ holds, giving the stated value of $\Theta$ . But this does not depend on the choices of $f,f',n,n'$ , so we have (i). Since the functions $\Delta_f$ have neither zeroes nor poles on $\Omega$ , (ii) results. Finally, (iii) comes out if one evaluates at a cusp.

5.16 Remark. Comparison of (5.15) with (IV 4.8, 4.10) shows $\Delta$ to be a two-dimensional analogue of the lattice invariant $\xi(\Lambda)^{q^\delta-1}$ . Let

$\phi$ be a 2-D-module with associated lattice $\Lambda$ and $\xi(\Lambda) = (q^{2\delta}-1)$-st root of $\Delta(\Lambda)$ . Then $\phi' = \xi \circ \phi \circ \xi^{-1}$ has its leading coefficients in $k^*$ . If the second exterior power $\Lambda^2(\Lambda)$ is isomorphic with $\mathfrak{d}$ , the mapping $a \longmapsto 1(\phi'_a)$ from A to $k$ agrees with the twisted sign function $\tau^{2-2g-\deg \mathfrak{d}}(\mathrm{sgn}(a))$ .

We are finishing the chapter with some examples for (5.11) resp. (5.12).

5.17 Examples. (i) Let $g = 0$ and $\delta \geq 1$ arbitrary. Then $\deg : \mathrm{Pic}\, A \xrightarrow{\cong} \mathbb{Z}/\delta$ . The partial Z-function $Z_{(\mathfrak{a})}(S)$ for an ideal $\mathfrak{a}$ of degree $j$ , where $0 \leq j < \delta$ , is given by

$$Z_{(j)} = Z_{(\mathfrak{a})} = (q-1)^{-1}[(q^{j+1}-1)S^j + q^{j+1}(q^\delta-1)S^{\delta+j}/(1-q^\delta S^\delta)] .$$

If $0 < i \leq \delta$ , put $\Delta_i = \Delta_{\mathfrak{n}}$ , where $\mathfrak{n}$ is integral of degree $i$ . On the principal component $\bar{M}_\Gamma$ , where $\Gamma = GL(2,A)$ , we have the cusps $s(j,k)$ , $j$ and $k$ satisfying $j+k \equiv 0(\delta)$ . The order $\mathrm{ord}(i,j)$ of $\Delta_i$ at $s(j,-j)$ is given for $\delta = 1,2,3$ by the matrices

$$(1) , \begin{pmatrix} q , & 1 \\ q^3+1 , & q^2+q \end{pmatrix} , \begin{pmatrix} q & 1 & q^2 \\ q^3+q^2 , & q^2+q , & q^4+1 \\ q^5+q^4+1 , & q^4+q^3+q^2 , & q^6+q^2+q \end{pmatrix} .$$

Note $0 < i \leq \delta$ , but $0 \leq j < \delta$ !

(ii) Let now $g = 1$ and $\delta = 1$ , and put $P(X) = qX^2-tX+1$ . In this case, $\mathrm{Pic}\, A - \{1\}$ is represented by the prime ideals $\mathfrak{p}$ of degree 1. We consider the following $h$ modular forms of weight $q^2-1$ for $\Gamma = GL(2,A)$ : $\{\Delta_\mathfrak{p} \mid \deg \mathfrak{p} = 1\} \cup \{\Delta\}$ . The partial Z-functions are given by

$$Z_{(\mathfrak{p})}(S) = S + qS^2/(1-qS)$$

$$Z_{(1)}(S) = 1 + qS^2/(1-qS) .$$

The order of $\Delta_\mathfrak{p}$ at the cusp $s(\mathfrak{b}^{-1},\mathfrak{b})$ is $Z_{(\mathfrak{b}\mathfrak{p})}(q)-q^2Z_{(\mathfrak{b})}(q)$ which equals

$$\begin{cases} q & (\mathfrak{h}) \neq 1 \neq (\mathfrak{h}\mathfrak{p}) \\ 1 & (\mathfrak{h}\mathfrak{p}) = 1 \\ q^3-q^2+q & (\mathfrak{h}) = 1 \end{cases}$$

respectively. The order of $\Delta$ at $s(\mathfrak{h}^{-1},\mathfrak{h})$ is $(1-q^2)Z_{(\mathfrak{h})}(q)$

$$= \begin{cases} q & (\mathfrak{h}) \neq 1 \\ q^3-q^2+1 & (\mathfrak{h}) = 1 \end{cases}.$$

Up to sign, the associated matrix has determinant $(q-1)^{h-1}P(q)$ .

In all the cases considered in (5.17), we know a priori $\bar{M}_\Gamma$ to have genus zero (see (5.9)), so all the cuspidal divisors of degree zero are principal, and the above described construction does not give all the principal divisors supported by the cusps.

## VII Modular Forms and Functions

In the whole chapter, $\Gamma$ is the group $GL(Y)$ of a 2-lattice $Y \subset K^2$ , and $\Gamma(\mathfrak{n}) \subset \Gamma$ the $\mathfrak{n}$-th congruence subgroup for an ideal $\mathfrak{n} \subsetneq A$ .

### 1. The Field of Modular Functions

We fix some non-constant element $a \in A$ of degree $d$ , and consider the modular forms $l_i(\omega) = l_i(a,\omega)$ of (V 3.4) . Between these forms and the Eisenstein series $E^{(j)}(\omega) = E^{(j)}(Y_\omega)$ , we have the relations ($k \geq 0$ arbitrary)

(1.1) $$a\,E^{(q^k-1)} = \sum_{i+j=k} E^{(q^i-1)} l_j^{q^i} ,$$

where we have put $E^{(0)} = -1$ (II 2.11).

1.2 Conclusion. The 2-lattice $Y_\omega$ resp. the associated D-module is completely determined by the values of

a) $E^{(q^k-1)}$ $\qquad k \in \mathbb{N}$ , or

b) $l_k$ $\quad 1 \leq k \leq 2d$ , or

c) $E^{(q^k-1)}$ $\quad 1 \leq k \leq 2d$ .

For any set $\{f\}$ of meromorphic modular forms, let $C(f)_0$ be the field of elements of weight 0 in $C(f)$ , i.e. the field of modular functions.

1.3 Proposition. The function field $C(M_\Gamma)$ of $M_\Gamma = \Gamma\backslash\Omega$ is $C(E^{(q^i-1)}\,|\,1 \leq i \leq 2d)_0 = C(l_i)_0$ .

Proof. It suffices to show:

"If $\omega,\omega' \in \Omega$ are such that for all $h \in C(l_i)_0$ , we have $h(\omega) = h(\omega')$ then $\omega$ and $\omega'$ are $\Gamma$-equivalent", resp.
"If $\Lambda,\Lambda'$ are 2-lattices in $C$ (isomorphic with $Y$ as A-modules) such that for all $h$ , we have $h(\Lambda) = h(\Lambda')$ , then $\Lambda' = \text{const.}\ \Lambda$ ."

We apply (IV 4.2) to $x_i = l_i(\Lambda')/l_i(\Lambda)$ , assigning the weight $q^i-1$ to $x_i$ $(1 \leq i \leq 2d)$ . By (IV 4.1), gcd $(q^i-1) = q-1$ , so $x_i = y^{(q^i-1)/(q-1)}$ for some $y \in C^*$ , and $\Lambda' = z\Lambda$ if $z^{q-1} = y$ . □

In the next step, we describe the function field $C(M_{\Gamma(n)})$ .

1.4 Lemma. The group of the ramified covering $\Gamma(n)\backslash\Omega \to \Gamma\backslash\Omega$ is $\{\gamma \in GL(Y/nY)\,|\det\gamma \in \mathbb{F}_q^*\}/Z(\mathbb{F}_q)$, $Z(\mathbb{F}_q)$ denoting the $\mathbb{F}_q$-valued scalar matrices.

Proof. By the strong approximation theorem for $SL(2)$ , the reduction mapping $SL(Y) \to SL(Y/nY)$ is surjective. Thus, the image of $\Gamma$ in $GL(Y/nY)$ consists of the elements of constant determinant. □

We have to construct some basic modular functions for $\Gamma(n)$ . For $u \in n^{-1}Y-Y$ , let

$$\begin{aligned} f_u(\omega) &= E^{(q-1)}(\omega)e_u^{q-1}(\omega) \\ &= (a^q-a)^{-1}l_1(\omega)e_u^{q-1}(\omega) . \end{aligned}$$

By the properties of the $e_u$ , we obtain:

(1.5) (i) If $\gamma \in \Gamma$ , then $f_u(\gamma\omega) = f_{u\gamma}(\omega)$ ;

(ii) $f_u = f_{u'} \Longleftrightarrow u' = c\cdot u$ with some $c \in \mathbb{F}_q^*$ .

If we have $f_u \circ \gamma = f_u$ for all the $u$ , we must have $u\gamma = c(u)\cdot u$ with $c(u) \in \mathbb{F}_q^*$ which cannot depend on $u$ , as one easily sees. Thus, $\gamma$ lies in $\Gamma(\mathfrak{n}) \cdot Z(\mathbb{F}_q)$ , and the following proposition holds.

1.6 Proposition. $C(M_{\Gamma(\mathfrak{n})})$ is generated over $C(M_\Gamma)$ by the functions $f_u$ .

In contrast with the Eisenstein series which are defined only above $C$ , the $l_i$ make sense for Drinfeld modules over arbitrary A-schemes. They are algebraic modular forms as defined in [27] (see also [41, App.1]) . Of course, the same is true for the modular form $\Delta$ defined in (5.14). We may therefore use the $l_i$ and the $f_u$ to determine the function field "over $K$ " of our modular schemes $M(1)$ and $M(\mathfrak{n})$. But first, we have to relate the $f_u$ with level $\mathfrak{n}$ structures.

(1.7) Let now $\mathfrak{n}$ be an admissible positive divisor, $M(\mathfrak{n})$ the modular scheme for 2-D-modules with a level $\mathfrak{n}$ structure, and $\phi$ the universal 2-D-module over $M(\mathfrak{n})$ with level structure $\alpha : (\mathfrak{n}^{-1}/A)^2 \to D(\phi,\mathfrak{n})$ . After the base change $\times_A K$ , $\alpha$ becomes an isomorphism of A-modules.

For each point $x$ of $M(\mathfrak{n}) \times K$ , $\phi$ induces a D-module $\phi^{(x)}$ with level $\mathfrak{n}$ structure $\alpha^{(x)}$ . If $\mathcal{L}$ denotes the line bundle over $M(\mathfrak{n}) \times K$ underlying $\phi$ , an element $u$ of $(\mathfrak{n}^{-1}/A)^2$ defines by means of $\alpha$ a section $e_u$ of $\mathcal{L}$ . Accordingly, the coefficients $l_i$ of $\phi_a$ define sections of $\mathcal{L}^{1-q^i}$ . In particular, we may consider the elements of weight zero

$$f_u = (a^q-a)^{-1} l_1\, e_u^{q-1} .$$

These lie in the function field $K(M(\mathfrak{n}))$ of $M(\mathfrak{n})$ and do not depend on the choice of the non-constant element $a$ . As a direct consequence of the definition (I 3), the Galois group

$$\mathrm{Gal}(M(\mathfrak{n}) : M(1)) = GL(2,A/\mathfrak{n})/Z(\mathbb{F}_q)$$

acts from the right on $f_u$ by

(1.8) $$f_u^\gamma = f_{u\gamma} .$$

This implies for divisors $\mathfrak{m}$ of $\mathfrak{n}$ :

$$f_u \in K(M(\mathfrak{m})) \iff u \in (\mathfrak{m}^{-1}/A)^2 .$$

Correspondingly, one derives: The definition of the forms resp. functions $l_i, e_u, f_u$ does not depend on the denominator $\mathfrak{n}$ used. After base change with $C$, we get back the modular forms defined earlier. But one should take care of the change of index in the $e_u$ !

1.9 Theorem. (i) The function field $K(M(1))$ of $M(1)$ is $K(l_i)_o$ .

(ii) The algebraic closure of $K$ in $K(M(1))$ is (isomorphic with) the field $H$ . The scheme $(\bar{M}(1)-M(1)) \times K$ is the disjoint union of points isomorphic with $\operatorname{Spec} H$ .

(iii) $K(M(\mathfrak{n}))$ is generated over $K(M(1))$ by the $f_u$ , $u$ running through $(\mathfrak{n}^{-1}/A)^2$ .

(iv) The algebraic closure of $K$ in $K(M(\mathfrak{n}))$ is the field $H(\mathfrak{n})$ . The scheme $(\bar{M}(\mathfrak{n})-M(\mathfrak{n})) \times K$ is the disjoint union of points isomorphic with $\operatorname{Spec} H(\mathfrak{n})$ .

Proof. Let $L$ be the algebraic closure of $K$ in $K(M(1))$ and $L^s$ its maximal separable subextension. The K-imbeddings $\sigma$ of $L^s$ into $C$ correspond to the components of $M(1) \underset{K}{\times} C = \coprod_{\sigma} M_\sigma$ , where $M_\sigma = M(1) \underset{L^s,\sigma}{\times} C$ . In particular, $L^s$ and $H$ have the same degree over $K$ . Now, $M(1) \times C$ being reduced, $L^s = L$ .

Let $l_{i,\sigma}$ be the restriction of $l_i$ to $M_\sigma$ . By (1.3),

$$K(M(1)) \underset{L,\sigma}{\otimes} C = C(M_\sigma) = C(l_{i,\sigma})_0 = L(l_i)_0 \underset{L,\sigma}{\otimes} C ,$$

i.e. $K(M(1)) = L(l_i)_0$ . To show (i), it will suffice to find a subfield of $K(l_i)_0$ isomorphic with $H$ . First, we note: Using (1.8), (iii) may be proved in the same way as (1.6). Next, we conclude as above: The algebraic closure $L(\mathfrak{n})$ of $K$ in $K(M(\mathfrak{n}))$ is a separable field extension of the same degree as $H(\mathfrak{n})$ .

Let $\mathfrak{n}$ be admissible. Choosing an embedding $\sigma : L(\mathfrak{n}) \hookrightarrow C$ and denoting $M(\mathfrak{n})_\sigma = M(\mathfrak{n}) \underset{L(\mathfrak{n}),\sigma}{\times} C$ etc., we have

$$K(M(\mathfrak{n})) \underset{L(\mathfrak{n}),\sigma}{\otimes} C = C(l_{i,\sigma}, f_{u,\sigma}) = L(\mathfrak{n})(l_i, f_u)_0 \underset{L(\mathfrak{n}),\sigma}{\otimes} C .$$

The universal 2-D-module $\phi$ above $M_\sigma$ with its $\mathfrak{n}$-structure and satisfying $\phi_a = \sum l_{i,\sigma} \tau^i$ degenerates at a cusp $s$ to a D-module of rank 1, and "evaluation at $s$" defines a valuation on $K(M(\mathfrak{n}))$ with residue class field $H(\mathfrak{n})$ (see [27, 1.78]). This shows (iv) for our admissible divisor $\mathfrak{n}$. By Galois descent, we obtain (iv) for arbitrary integral divisors as well as (ii). We are left to show $H \subset K(l_i)_0$. For this, it will suffice to have $H(\mathfrak{n}) \subset K(l_i, f_u)_0$ for each admissible $\mathfrak{n}$.

Now the D-module $\phi$ corresponds to some

$$\beta : \operatorname{Spec} K[l_i, e_u] \to M(\mathfrak{n}) \times K$$

whose dual morphism $\beta^*$ maps the constants $H(\mathfrak{n})$ to $K(l_i, e_u)_0$. The isomorphism class of $\phi$ will not be changed by $e_u \longmapsto e_{cu} = ce_u$ $(c \in \mathbb{F}_q^*)$, so $\beta^*(H(\mathfrak{n}))$ is contained in $K(l_i, e_u)_0^{\mathbb{F}_q^*} = K(l_i, f_u)_0$. □

## 2. The Field of Definition of the Elliptic Points

In this section, $\delta$ is odd. Let $K' = K\mathbb{F}_{q^2} \subset C$ with ring of integers $A'$ and Hilbert class field $H'$. Then $H'$ is Galois over $K$, and we have an exact sequence

$$(2.1) \qquad 1 \to \operatorname{Gal}(H' : K') \to \operatorname{Gal}(H' : K) \to \mathbb{Z}/2 \to 1 .$$
$$\qquad \operatorname{Gal}(H' : K') \cong \operatorname{Pic} A'$$

The non-trivial automorphism of $K'_\infty$ over $K_\infty$ induces a conjugation $\beta$ of $H'$ whose fixed field we denote by $H_\beta$.

Let $\mathfrak{n}$ be an admissible divisor of $A$. After having fixed an isomorphism $\mathfrak{n}^{-1}A'/A' \xrightarrow{\cong} (\mathfrak{n}^{-1}/A)^2$, there results a morphism

$$M^1_{A'}(\mathfrak{n}) \longrightarrow M^2(\mathfrak{n})$$
$$\phi \longmapsto \phi|_A$$

which associates the 2-D-module $\phi|_A$ to each 1-D-module $\phi$ for $A'$ with a structure of level $\mathfrak{n}A'$. (Here $M^1_{A'}(\mathfrak{n})$ denotes the modular scheme of 1-D-modules for the Dedekind ring $A'$ playing the part of $A$.)

By Galois descent, we get a morphism

$$M^1_{A'}(1) \longrightarrow M^2(1) ,$$

finally, after applying $\times_A K$, an $H'$-valued point

$$\varepsilon : \mathrm{Spec}\, H' \longrightarrow M^2(1) \times K \qquad \text{(I 4.1)} .$$

The image of the base change $\varepsilon \times_K C$ consists precisely of the elliptic points of $M^2(1) \times C$.

For each $\sigma \in \mathrm{Gal}(H' : K)$, there is an elliptic point $e_\sigma$, and we have

$$e_{\sigma'} = e_\sigma \Longleftrightarrow \sigma' = \sigma \quad \text{or} \quad \sigma' = \sigma\beta .$$

In fact, if $\phi$ is a 1-D-module for $A'$ having its coefficients in $H'$, then $\phi|_A$ is $C$-isomorphic with $\beta(\phi|_A)$. Conversely, $\phi|_A \cong \phi'|_A$ implies $\phi \cong \phi'$. This shows

2.2 Proposition. (i) The image of $\varepsilon$ is an $H_\beta$-valued point of $M^2(1) \times K$.

(ii) All the elliptic points of $M^2(1) \times C$ are $\mathrm{Gal}(\bar{K} : K)$ - conjugate.

## 3. Behavior of $E^{(q-1)}$ at Elliptic Points

Again we assume $\delta$ to be odd. We are interested in the behavior of the Eisenstein series $E^{(q-1)}$ at the elliptic points of $M(1) \times C$. Clearly, each $f \in M_{q-1}(\Gamma)$ vanishes at elliptic points, but we need to know the precise order of vanishing.

We first compute the order of

$$E(\omega) = \sum_{u,v \in A}{}' \, (u\omega+v)^{1-q}$$

in the elliptic point $e \in \mathbb{F}_{q^2} - \mathbb{F}_q$ of $\Gamma\backslash\Omega$ , where $\Gamma = GL(2,A)$ .

Note: For $x,y \in K$ ,

(3.1) $$|x+ey| = \sup(|x|,|y|)$$

holds. For the derivative of $E$ , we have

(3.2) $$E'(\omega) = \sum_{u,v}{}' u(u\omega+v)^{-q}$$

$$= \sum_{u}{}' u\, e_A^{-q}(u\omega) \qquad \text{by (I 2.2v)} .$$

Further, in view of (3.1),

$$\sum_{u \in \mathbb{F}_q^*} u\, e_A^{-q}(ue) = \sum u^{1-q} e_A^{-q}(e)$$

$$= - e_A^{-q}(e)$$

$$= - e^{-q} \prod_{v \in A}{}' (1-e/v)^{-q}$$

has absolute value 1. Considering $u \in A$ with $|u| > 1$ , the term $u(ue+v)^{-q}$ has absolute value

$$|u|\,|ue+v|^{-q} = |u|\sup(|u|,|v|)^{-q} \leqq |u|^{1-q} < 1 .$$

Hence, in the sum (3.2) for $E'(e)$ , the contribution of those $(u,v)$ with $u$. constant is dominating, i.e. $E'(e) \neq 0$ , and $E$ has a simple zero at $e$ .

3.3 <u>Proposition</u>. $E^{(q-1)}$ has simple zeroes at all the elliptic points.

<u>Proof</u>. $E^{(q-1)}$ comes from the algebraic modular form $(a^q-a)^{-1} l_1(\phi_a)$ , where $a \in A$ is arbitrary, but non-constant. Now $l_1$ is defined over $K$ , and the zero order is Galois-invariant, so the assertion follows from (2.2) and $E'(e) \neq 0$ .

## 4. The Graded Algebra of Modular Forms

We will determine the C-algebra $\bigoplus_{k \geq 0} M_k(\Gamma)$ of modular forms for $\Gamma$. By (VI 5.10), we already know the dimension of $M_k(\Gamma)$, provided that $k \equiv 0(q^2-1)$.

Let $0 \neq f \in M_k(\Gamma)$. The divisor $\operatorname{div}(f)$ of $f$ is defined to be the formal linear combination

$$(4.1) \qquad \operatorname{div}(f) = \sum_{x \in \bar{M}_\Gamma} n_x \cdot x ,$$

where $n_x$ is the order of $f$ at $x$, if $x$ is not elliptic, and the order of $f$ at $y$ divided by $(q+1)$, if $x$ is represented by the elliptic point $y \in \Omega$.

Let further $[D]$ be the integral part of the divisor $D$ with rational coefficients, and $\deg D \in \mathbb{Q}$ the sum of its coefficients.

Denoting by $E$ the Eisenstein series $E^{(q-1)}$, we obviously have: The mapping $f \longmapsto f/E$ is an isomorphism of $M_{q-1}(\Gamma)$ with the space $H^0(\bar{M}_\Gamma, [\operatorname{div}(E)])$ of modular functions whose divisors are $\geq -[\operatorname{div}(E)]$. Accordingly, for all $k \in \mathbb{N}$,

$$(4.2) \qquad M_{k(q-1)}(\Gamma) \xrightarrow{\cong} H^0(\bar{M}_\Gamma, [k \cdot \operatorname{div}(E)])$$

$$f \longmapsto f/E^k .$$

Using (VI 5.7), we obtain

$$(4.3) \qquad \deg \operatorname{div}(E) = \frac{q^\delta - 1}{q^2-1} P(q) .$$

If now $\delta$ is even, there are no elliptic points and $\operatorname{div}(E)$ has integral coefficients. For an odd $\delta$, by means of (3.3) and (V 4.5):

$$(4.4) \qquad \deg[k \cdot \operatorname{div}(E)] = k\left(\frac{q^\delta-1}{q^2-1}\right) P(q) - \langle k/(q+1)\rangle P(-1) ,$$

writing $\langle n\rangle$ for the fractional part of a rational number $n$. (There should be no confusion with the use of $\langle\ \rangle$ in Chapter IV.)

A trivial estimation implies

(4.5) $\deg[\operatorname{div}(E)] \geq 2g(\bar{M}_\Gamma)-1$ ,

provided that $q > 2$ . In this case, we are able to compute $M_{k(q-1)}(\Gamma)$ by Riemann-Roch. However, if $q = 2$ , we have to assume $k \geq 2$ to get a formula for $\dim M_{k(q-1)}(\Gamma)$ ,

$$\deg[\operatorname{div}(E)] \geq g(\bar{M}_\Gamma)$$

being here the best generally valid estimation. Taken together, we obtain the formula, valid in all cases except eventually $(q,k) = (2,1)$:

(4.6) $$\dim M_{k(q-1)}(\Gamma) = 1-g(\bar{M}_\Gamma)+k(\frac{q^\delta-1}{q^2-1})P(q) ,$$

minus $\langle k/(q+1)\rangle P(-1)$ in case $\delta$ is odd.

4.7 <u>Examples.</u> <u>$(g,\delta) = (0,1)$</u> : see (V 3.6).
<u>$(g,\delta) = (0,2)$</u> : We have $g(\bar{M}_\Gamma) = 0$ and $\dim M_{q-1}(\Gamma) = 2$ . The vector space $M_{q-1}(\Gamma)$ is spanned by $E$ and $\tilde{E}$ , where $\tilde{E}(\omega) = E^{(q-1)}(\mathfrak{p}^{-1}Y_\omega)$ , $\mathfrak{p}$ a prime ideal of degree 1. The quotient $\tilde{E}/E$ is a parameter for the projective line $\bar{M}_\Gamma$ .

<u>$(g,\delta) = (0,3)$</u> : Again, $g(\bar{M}_\Gamma) = 0$ , but $\dim M_{q-1}(\Gamma) = q+1$ . Each $f \in M_{q-1}(\Gamma)$ has a zero (counted with multiplicity $1/(q+1)$ ) at the elliptic point, and $q$ further zeroes. Problem: Give a geometric description of a basis $\{f\}$ of $M_{q-1}$ , and of the zeroes of the functions $f$ !

<u>$(g,\delta) = (1,1)$</u> : As in the other examples, $g(\bar{M}_\Gamma) = 0$ . Putting $P(X) = qX^2 - tX+1$ , we have $\dim M_{q-1}(\Gamma) = q^2-q+1-t$ . As above, I do not know a "canonical" basis.

## 5. Higher Modular Curves

In order to save unnecessary efforts of writing and notation, we now restrict the discussion to the principal component of $\bar{M}(1) \times C$ . However, without too much difficulties, the results of this section may be transferred to the general case.

Thus let $\Gamma = GL(2,A)$ . For the determination of the genera of higher modular curves, we need the Hurwitz formula.

(5.1) Let $M$ be a connected, projective, nonsingular algebraic curve over some algebraically closed field $C$ of characteristic $p > 0$, and let $G$ be a finite group of automorphisms of $M$ with quotient curve $N = M/G$.

Further, for a C-valued point $x$ of $M$, let $\mathcal{O}_x$ be the local ring, $\pi_x$ a uniformizing parameter at $x$ and $G_{x,i}$ the i-th ramification group. Then $G_x = G_{x,0} \supset G_{x,1} \supset \ldots G_{x,r} = 1$ if $r$ is sufficiently large,

$$G_{x,i} = \{\gamma \in G | \gamma \text{ acts trivially modulo } \pi_x^{i+1}\}.$$

For an element $\gamma \neq 1$ of $G_x$, put

$$i_x(\gamma) = \sup\{i | \gamma \in G_{x,i}\} + 1 .$$

The Euler numbers $e(\ldots) = 2-2g(\ldots)$ of $M$ and $N$ are then related by the Hurwitz formula

(5.2) $$e(M) = \#(G) \cdot e(N) - \sum j_x ,$$

summing over the ramified points $x$ of $M$. The contributions $j_x$ are given by

$$j_x = \sum_{1 \neq \gamma} i_x(\gamma) ,$$

see [67], or [58].

The groups $G_{x,i}$ are p-groups if $i \geq 1$, whereas the order of $G_x/G_{x,1}$ is prime to $p$. The (ramified) point $x$ is called wildly ramified, if $G_{x,1} \neq 1$, and tamely ramified otherwise.

We are going to investigate the ramification of $\bar{M}_{\Gamma(\mathfrak{n})}$ over $\bar{M}_\Gamma$, where $\mathfrak{n} \subset A$ has degree $d$. Let

$$G(\mathfrak{n}) = \Gamma/\Gamma(\mathfrak{n})Z(\mathbb{F}_q) = \{\gamma \in GL(2,A/\mathfrak{n}) | \det \gamma \in \mathbb{F}_q^*\}/Z(\mathbb{F}_q)$$

be the covering group. The following facts are obvious:

(5.3) $M_{\Gamma(\mathfrak{n})} : M_\Gamma$ ramifies precisely in the elliptic points. These points are tamely ramified with ramification index $q+1$.

(5.4) $Sp(\Gamma) = \Gamma \setminus \mathbb{P}_1(K)$ , accordingly for $\Gamma(\mathfrak{n})$ .

Recall: The cusp "$\infty$" occurring in Ch.VI corresponds to $(1:0) \in \mathbb{P}_1(K)$ .

(5.5) The stabilizer of $\infty \in \bar{M}_{\Gamma(\mathfrak{n})}$ is

$$G(\mathfrak{n})_\infty = \left\{\begin{pmatrix} a & b \\ 0 & d \end{pmatrix} \middle| a,d \in \mathbb{F}_q^*\right\}/Z(\mathbb{F}_q)$$

which has the order $(q-1)q^d$ .

5.6 Lemma. The stabilizers of all cusps of $\bar{M}_{\Gamma(\mathfrak{n})}$ are conjugate in $G(\mathfrak{n})$ to $G(\mathfrak{n})_\infty$ .

Proof. Let $(a:c) \in \mathbb{P}_1(K)$ , where $a$ and $c$ are non-zero elements of $A$ , and suppose $(a) = \mathfrak{a} \cdot \mathfrak{g}$ , $(c) = \mathfrak{c} \cdot \mathfrak{g}$ with relatively prime positive divisors $\mathfrak{a}$ and $\mathfrak{c}$ . Put $\gamma = \begin{pmatrix} a & 0 \\ c & 1 \end{pmatrix}$ . We will compute $\Gamma \cap B(K)^\gamma$ , $B$ denoting here the group of upper triangular matrices. The relation

$$(*) \qquad \gamma \begin{pmatrix} r & s \\ 0 & t \end{pmatrix} \gamma^{-1} \in \Gamma$$

implies $r,t \in \mathbb{F}_q^*$ , $cs \in \mathfrak{c}$ and $cs \equiv r-t \bmod \mathfrak{a}$ . Conversely, to arrive at (*), we may arbitrarily choose $r,s,t$ satisfying these conditions. This shows

$$\Gamma \cap B(K)^\gamma \pmod{\Gamma(\mathfrak{c})Z(\mathbb{F}_q)} = G(\mathfrak{c})_\infty$$

and

$$\Gamma \cap B(K)^\gamma \pmod{\Gamma(\mathfrak{n})Z(\mathbb{F}_q)} = G(\mathfrak{n})_\infty$$

in case $\mathfrak{n}$ divides $\mathfrak{c}$ .

Each element of Pic A may be represented by an ideal $\mathfrak{g}$ as above. Thus, over each cusp of $\bar{M}_\Gamma$ , we find one of $\bar{M}_{\Gamma(\mathfrak{n})}$ having $G(\mathfrak{n})_\infty$ as its stabilizer. □

5.7 Lemma. The higher ramification groups $G(\mathfrak{n})_{s,i}$ of cusps $s$ of $\bar{M}_{\Gamma(\mathfrak{n})}$ satisfy $\#(G(\mathfrak{n})_{s,1}) = q^d$ and $G(\mathfrak{n})_{s,2} = 1$ .

Proof [17, 3.4.7]. The first assertion is obvious by divisibility reasons. As to the second, we have to show: For each non-trivial $\sigma \in G(\mathfrak{n})_{s,1}$ , $\sigma(t_s)-t_s$ has a zero of order 2 at $s$ ( $t_s$ denoting a parameter at $s$ ). Without restriction, let $s = \infty$ . Around $\infty$ , the projection $\bar{M}_{\Gamma(\mathfrak{n})} \to \bar{M}_\Gamma$ is isomorphic as a ramified covering of analytic manifolds with $\mathfrak{n}\backslash(\Omega \cup \{\infty\}) \to \Gamma_\infty\backslash(\Omega \cup \{\infty\})$ . The group

$$A/\mathfrak{n} \xrightarrow{\cong} G(\mathfrak{n})_{\infty,1}$$

$$b \longmapsto \begin{pmatrix} 1 & b \\ 0 & 1 \end{pmatrix} = \sigma_b$$

acts by translation: $\sigma_b(z) = z+b \bmod \mathfrak{n}$ . For small values of the parameter $t = e_\mathfrak{n}^{-1}(z)$ and $b \in A-\mathfrak{n}$ , we have

$$\begin{aligned} \sigma_b(t)-t &= e_\mathfrak{n}^{-1}(z+b) - e_\mathfrak{n}^{-1}(z) \\ &= (e_\mathfrak{n}(z)+e_\mathfrak{n}(b))^{-1} - e_\mathfrak{n}^{-1}(z) \\ &= e_\mathfrak{n}^{-2}(z)(-e_\mathfrak{n}(b)+e_\mathfrak{n}^2(b)e_\mathfrak{n}^{-1}(z)-\ldots) \\ &= t^2 \cdot \text{unit in } t . \qquad \square \end{aligned}$$

We are now able to determine $g(\bar{M}_{\Gamma(\mathfrak{n})})$ , having collected all the ingredients of the Hurwitz formula:

(5.8) Number of cusps of $\bar{M}_{\Gamma(\mathfrak{n})} = h \cdot \#(G(\mathfrak{n})/G(\mathfrak{n})_\infty$

$$= h(\mathfrak{n}) \ ;$$

(5.9) Contribution $j_s$ of the cusps:

$$j_s = \sum_{1 \neq \gamma \in G(\mathfrak{n})_s} i_s(\gamma) = \#(G(\mathfrak{n})_\infty - G(\mathfrak{n})_{\infty,1}) + 2\#(G(\mathfrak{n})_{\infty,1}-1)$$

$$= q^{d+1} - 2 \ ;$$

(5.10) Number of points $x$ of $\bar{M}_{\Gamma(\mathfrak{n})}$ above elliptic points of $\bar{M}_\Gamma$

$$= \begin{cases} 0 & (\delta \text{ even}) \\ P(-1)\#(G(\mathfrak{n}))/(q+1) & (\delta \text{ odd}) , \end{cases}$$

each with $j_x = q$ .

Combining (5.2) and (VI 5.8), we obtain by an elementary computation which will be omitted:

5.11 Theorem. The genus of $\bar{M}_{\Gamma(\mathfrak{n})}$ is given by

$$1 + \frac{q^{\delta}-1}{(q^2-1)(q-1)} P(q) \#(G(\mathfrak{n})) - h(\mathfrak{n}) .$$

The congruence subgroup $\Gamma(\mathfrak{n})$ being p'-torsion free, the conditions of Thm.13/14 of [61] are fulfilled, and the formula given there agrees with ours.

Having done all the preparations, it is a purely combinatorial problem to determine the genus of $\bar{M}_{\Gamma'}$, $\Gamma'$ being now an arbitrary congruence subgroup. From an arithmetical point of view, the most interesting curves are the curves $\bar{M}_{\Gamma_0(\mathfrak{n})}$ associated with Hecke congruence subgroups

$$\Gamma_0(\mathfrak{n}) = \left\{\begin{pmatrix} a & b \\ c & d \end{pmatrix} \in \Gamma \middle| c \equiv 0 \bmod \mathfrak{n}\right\} .$$

Their genera are best obtained by means of the covering $\bar{M}_{\Gamma(\mathfrak{n})} \to \bar{M}_{\Gamma_0(\mathfrak{n})}$. In this covering, elliptic ramification occurs if and only if $\delta$ is odd and, additionally, all the prime divisors of $\mathfrak{n}$ are of even degree [17, 3.4.17]. A cusp $s$ of $\Gamma(\mathfrak{n})$ having $G(\mathfrak{n})^{\gamma}_{\infty}$ as its stabilizer will contribute to the Hurwitz formula by

$$(5.12) \qquad j_s = \#(B(\mathfrak{n}) \cap G(\mathfrak{n})^{\gamma}_{\infty}) + \#(B(\mathfrak{n}) \cap G(\mathfrak{n})^{\gamma}_{\infty,1}) - 2 ,$$

$B(\mathfrak{n})$ denoting here the group of upper triangular matrices in $G(\mathfrak{n})$. The evaluation of this formula, depending on the prime decomposition of $\mathfrak{n}$, is rather complicated, but carried out in full detail in [17]. There, only the case $K = \mathbb{F}_q(T)$ of a rational function field is treated; nevertheless, (5.6) implies the validity of the formula [17, 3.4.15] which gives the contribution $\sum j_s$ of all cusps $s$ of $\Gamma(\mathfrak{n})$ above one fixed cusp of $\Gamma$, in the most general case.

We will write down the result only in the case of a prime ideal.

5.13 Theorem. Let $\mathfrak{p}$ be a prime ideal of degree $d$. The genus of $\bar{M}_{\Gamma_0(\mathfrak{p})}$ is given by

$$g(\bar{M}_{\Gamma_o(\mathfrak{p})}) = 1+(q^2-1)^{-1}[\frac{(q^d+1)(q^\delta-1)}{q-1} P(q)-\delta q(q+1)P(1)+\eta] ,$$

where now $\eta = -q(q-1)P(-1)$ in case $d$ is even and $\delta$ is odd, and $\eta = 0$ otherwise.

We give a table of these numbers for the groups $\Gamma_0(\mathfrak{p})$ and $\Gamma(\mathfrak{p})$, where $\mathfrak{p}$ is a prime of degree $d$, and for small values of $g,\delta,d$. In the case $g = 1$, we put $P(X) = qX^2 - tX+1$.

5.14 Table.

| $g$ | $\delta$ | $d$ | $g(\bar{M}_{\Gamma_o(\mathfrak{p})})$ | $g(\bar{M}_{\Gamma(\mathfrak{p})})$ |
|---|---|---|---|---|
| 0 | 1 | 1 | $0$ | $0$ |
| | | 2 | $0$ | $q(q^3-q^2-1)$ |
| | | 3 | $q$ | $q^7 + \ldots$ |
| | | 4 | $q^2$ | $q^{10} + \ldots$ |
| | 2 | 1 | $0$ | $q^2-q-1$ |
| | | 2 | $q$ | $q^5 + \ldots$ |
| | | 3 | $q(q+1)$ | $q^8 + \ldots$ |
| | | 4 | $q(q^2+q+1)$ | $q^{11} + \ldots$ |
| | 3 | 1 | $q$ | $q^3+q^2-2q-2$ |
| | | 2 | $q(q+1)$ | $q^6 + \ldots$ |
| | | 3 | $q(q^2+q+2)$ | $q^9 + \ldots$ |
| 1 | 1 | 1 | $q^2$ | $q^4-q^2-q-t(q^2-q-1)$ |
| | | 2 | $q(q^2-t)$ | $q^7 + \ldots$ |
| | 2 | 1 | $q(q^2+2q-t)$ | $q^5 + \ldots$ |
| | | 2 | $q[q^3+q^2+2q+1-t(q+1)]$ | $q^8 + \ldots$ |

5.15 Remarks. (i) The formulae (5.11) and (5.13) are also valid for the non-principal components of the modular schemes $\bar{M}(\mathfrak{n}) \times C$ resp. $\bar{M}_o(\mathfrak{n}) \times C$, the proof being more or less identical.

(ii) If the characteristic $p$ is unequal to 2, the curve $\bar{M}_{\Gamma'}$ associated to $\Gamma' = SL(2,A)$ is a two-sheet covering of $\bar{M}_{\Gamma}$. A formula for its genus is given in [23].

## 6. Modular Forms for Congruence Subgroups

Up to some restrictions, we may compute $M_k(\Gamma')$ for congruence subgroups $\Gamma' \subset \Gamma$ by the method used in Section 4. Let first $\Gamma' = \Gamma(\mathfrak{n})$. The degree of the line bundle of modular forms of weight 1 for $\Gamma(\mathfrak{n})$ is

$$(6.1) \qquad \deg \mathfrak{M}_{1,\Gamma(\mathfrak{n})} = \frac{q^{\delta}-1}{(q-1)(q^2-1)} P(q) \#(G(\mathfrak{n})) \qquad (\text{see (VI } 5.7)) .$$

We remark this equals $(-1)$ times the Euler-Poincaré characteristic $\chi(\Gamma(\mathfrak{n}))$ of $\Gamma(\mathfrak{n})$ as defined in [60,61]. Comparison with the genus and an elementary estimation will give the inequality

$$(6.2) \qquad \deg \mathfrak{M}_{k,\Gamma(\mathfrak{n})} \geq 2g(\bar{M}_{\Gamma(\mathfrak{n})})-1$$

which holds for $k \geq 2$.

Therefore, Riemann-Roch suffices to compute the dimension of $M_k(\Gamma(\mathfrak{n}))$. However, the degree of $\mathfrak{M}_{1,\Gamma(\mathfrak{n})}$ is roughly about the genus, so we obtain only

$$(6.3) \qquad \dim M_1(\Gamma(\mathfrak{n})) = h(\mathfrak{n}) + \dim H^1(\bar{M}_{\Gamma(\mathfrak{n})}, \mathfrak{M}_1) ,$$

$h(\mathfrak{n}) = h \cdot \#(G(\mathfrak{n})/G(\mathfrak{n})_{\infty})$ denoting the number of cusps of $\Gamma(\mathfrak{n})$.

Let $\mathfrak{D}$ be the sheaf of holomorphic differentials on $\bar{M}_{\Gamma(\mathfrak{n})}$. Then

$$\begin{aligned} \dim H^1(\mathfrak{M}_1) &= \dim H^0(\mathfrak{D} \otimes \mathfrak{M}_1^{-1}) \\ &= \dim H^0(\mathfrak{M}_1^{oo}) \qquad \text{(V 5.7)} \\ &= \dim M_1^{oo}(\Gamma(\mathfrak{n})) , \end{aligned}$$

where we have written $\mathfrak{M}_1^{oo}$ for the sheaf of holomorphic modular forms of weight 1 having at least double zeroes at all the cusps, and $M_1^{oo}$ for its space of sections. In fact, one can find $h(\mathfrak{n})$ linearly independent Eisenstein series of weight 1 for $\Gamma(\mathfrak{n})$. It is not known whether there can exist non-zero elements in $M_1^{oo}(\Gamma(\mathfrak{n}))$ .

Let now $\Gamma' = \Gamma_o(\mathfrak{p})$ , $\mathfrak{p}$ a prime ideal of degree $d$ . We then have $M_k(\Gamma') = 0$ unless $k \equiv 0(q-1)$ . If $\delta$ is even or $d$ is odd, $\Gamma'$ has no elliptic fixed points and $M_k(\Gamma')$ is the space of sections of a line bundle; otherwise, we have to argue as in Section 4. In the latter case, there are precisely two elliptic points of $M_{\Gamma_o(\mathfrak{p})}$ above each elliptic point of $M_\Gamma$ [17, 3.4.17]. The result is

6.4 <u>Proposition</u>. Let $\mathfrak{p}$ be a prime ideal of degree $d$ . If $(q,k) \neq (2,1)$ , the dimension of $M_{k(q-1)}(\Gamma_o(\mathfrak{p}))$ is given by

$$\dim M_{k(q-1)}(\Gamma_o(\mathfrak{p})) = 1-g(\bar{M}_{\Gamma_o(\mathfrak{p})})+k\,\frac{(q^\delta-1)(q^d+1)}{(q^2-1)}\,P(q) ,$$

minus $2\langle k/(q+1)\rangle P(-1)$ in case $\delta$ is odd and $d$ is even.

If $(q,k) = (2,1)$ , the formula given is a lower bound for the dimension.

Some of these dimensions are given below. The notations are as in (5.14); if $q = 2$ , we suppose $k$ to be $\geq 2$ .

6.5 <u>Table</u>.

| g | $\delta$ | d | $\dim M_{k(q-1)}(\Gamma_o(\mathfrak{p}))$ |
|---|---|---|---|
| 0 | 1 | 1 | $1+k$ |
| | | 2 | $1+k(q^2+1)/(q+1)-2\langle k/(q+1)\rangle$ |
| | | 3 | $1-q+k(q^2-q+1)$ |
| | 2 | 1 | $1+k(q+1)$ |
| | | 2 | $1-q+k(q^2+1)$ |
| | 3 | 1 | $1-q+k(q^2+q+1)$ |
| 1 | 1 | 1 | $1-q^2+k\,P(q)$ |
| | | 2 | $1+q(q^2-t)+k(q^2+1)/(q+1)P(q)-2\langle k/(q+1)\rangle P(-1)$ |

# VIII Complements

## 1. Hecke Operators

From the operation of $GL(2,\mathbb{A}_f)$ on $\bar{M} = \bar{M}^2$, one derives the operation of a certain ring of correspondences on $\bar{M}(\mathfrak{n})$.

Let $\mathfrak{n}$ and $\mathfrak{m}$ be relatively prime ideals of $A$, and let $\alpha : (\mathfrak{n}^{-1}/A)^2 \to (\mathfrak{n}^{-1}\Lambda)/\Lambda$ be a $\mathfrak{n}$-structure of the 2-lattice $\Lambda$ in $C$. Let further $\Lambda'$ be a lattice containing $\Lambda$ such that the index $[\Lambda':\Lambda]$ is prime to $\mathfrak{n}$. The inclusion $\Lambda \subset \Lambda'$ induces a $\mathfrak{n}$-structure $\alpha'$ on $\Lambda'$. We define the following correspondence $T_\mathfrak{m}$ on the set of 2-lattices provided with a $\mathfrak{n}$-structure:

$$(1.1) \qquad T_\mathfrak{m}(\Lambda,\alpha) = \sum_{[\Lambda':\Lambda] = \mathfrak{m}} (\Lambda',\alpha'),$$

summing over all the lattices $\Lambda'$ that contain $\Lambda$ with index $\mathfrak{m}$. $T_\mathfrak{m}$ is called the $\mathfrak{m}$-th <u>Hecke operator</u>. We further put

$$S_\mathfrak{m}(\Lambda,\alpha) = (\Lambda',\alpha'),$$

where $\Lambda' = \mathfrak{m}^{-1}\Lambda$. Obviously, $S_{\mathfrak{m}\mathfrak{m}'} = S_\mathfrak{m}S_{\mathfrak{m}'}$.

1.2 <u>Lemma</u>. (i) For a prime ideal $\mathfrak{p}$ of degree $d$, there are precisely $q^d+1$ lattices $\Lambda' \supset \Lambda$ satisfying $[\Lambda':\Lambda] = \mathfrak{p}$, and

$$(T_{\mathfrak{p}^n})(T_\mathfrak{p}) = T_{\mathfrak{p}^{n+1}} + q^d T_{\mathfrak{p}^{n-1}} S_\mathfrak{p}$$

holds.

(ii) If $\mathfrak{m}$ and $\mathfrak{m}'$ are relatively prime, then $T_{\mathfrak{m}\mathfrak{m}'} = T_\mathfrak{m}T_{\mathfrak{m}'}$.

<u>Proof</u>. Easy, see e.g. [59,p.159]. But note the different definition of Hecke operators given there, using sublattices $\Lambda' \subset \Lambda$.

The free commutative algebra over $\mathbb{Z}$ generated by the $S_\mathfrak{m}, T_\mathfrak{m}$ $((\mathfrak{m},\mathfrak{n}) = 1)$ is called the <u>Hecke algebra</u> $\mathbb{H}$ (of level $\mathfrak{n}$).

1.3 <u>Remarks</u>. (i) $T_\mathfrak{m}$ acts on meromorphic modular forms $f$ of level $\mathfrak{n}$ by $T_\mathfrak{m}f(\Lambda,\alpha) = f(T_\mathfrak{m}(\Lambda,\alpha))$. If $f$ is holomorphic resp. a cusp form,

$T_{\mathfrak{m}}f$ will have the same property. For this, it is enough to see that $T_{\mathfrak{m}}f$ is bounded around the cusps (resp. $= 0$) if $f$ is.

(ii) Considering this action, we have for each prime ideal $\mathfrak{p}$

$$T_{\mathfrak{p}^i}f = (T_{\mathfrak{p}})^i f ,$$

the formal difference of both sides being divisible by $p$.

(iii) It is usual to regard Hecke operators as correspondences on the set of similarity classes of lattices (i.e. in our case: on the set of isomorphism classes of Drinfeld modules). Assuming this point of view, we may describe $T_{\mathfrak{m}}$ as the sum of the action of certain elements of $GL(2,\mathbb{A}_f)$ on $M(\mathfrak{n})$, see [69, Ch.VI]. In fact, $T_{\mathfrak{m}}$ induces an algebraic correspondence already on $M(\mathfrak{n})$ (i.e. not only on the set of C-valued points), and has a canonical extension to $\overline{M}(\mathfrak{n})$. Note however: $T_{\mathfrak{m}}$ does not respect the components of $M(\mathfrak{n})(C)$ !

(iv) Instead of considering full level $\mathfrak{n}$ structures, we may use coarser level structures and define $T_{\mathfrak{m}}$ analogously.

It is easy to find some eigenvectors for the operation of $H$ on modular forms.

1.4 Proposition. (i) The Eisenstein series $E^{(k)}$ satisfy $T_{\mathfrak{p}}E^{(k)}(\Lambda) = E^{(k)}(\mathfrak{p}^{-1}\Lambda)$. In particular, if $\mathfrak{p}$ is a principal ideal $(f)$, we have $T_{\mathfrak{p}}E^{(k)} = f^k E^{(k)}$.

(ii) For non-zero $u \in (\mathfrak{n}^{-1}/A)^2$, let

$$E_u^{(k)}(\Lambda,\alpha) = \sum_{\substack{\lambda \in \mathfrak{n}^{-1}\Lambda \\ \lambda \equiv \alpha(u) \bmod \Lambda}} \lambda^{-k}$$

the Eisenstein series of (V 3.7), the indices somewhat modified. Then $T_{\mathfrak{p}}E_u^{(k)}(\Lambda,\alpha) = E_u^{(k)}(\mathfrak{p}^{-1}\Lambda,\alpha')$.

Proof. $T_{\mathfrak{p}}E^{(k)}(\Lambda) = \sum_{[\Lambda':\Lambda]=\mathfrak{p}} \sum_{\lambda \in \Lambda'} \lambda^{-k}$

$$= \sum_{\lambda \in \mathfrak{p}^{-1}\Lambda}{}' m(\lambda)\lambda^{-k} ,$$

where
$$m(\lambda) = \#\{\Lambda' \mid \lambda \in \Lambda'\} = \begin{cases} q^d+1 & (\lambda \in \Lambda) \\ 1 & (\lambda \notin \Lambda) \end{cases}$$
$$\equiv 1(q) \quad , \quad \text{so}$$
$$T_{\mathfrak{p}} E^{(k)}(\Lambda) = E^{(k)}(\mathfrak{p}^{-1}\Lambda) \quad .$$

Accordingly,
$$T_{\mathfrak{p}} E_u^{(k)}(\Lambda,\alpha) = \sum_{\lambda \in \mathfrak{n}^{-1}\mathfrak{p}^{-1}\Lambda}{}' m(\lambda)\lambda^{-k} \text{ , where}$$
$$m(\Lambda) = \#\{\Lambda' \mid \lambda \equiv \alpha(u) \bmod \Lambda\}$$
$$= \begin{cases} 0 & \lambda \not\equiv \alpha(u) \mod \mathfrak{p}^{-1}\Lambda \\ 1 & \lambda \equiv \alpha(u) \mod \mathfrak{p}^{-1}\Lambda, \lambda \not\equiv \alpha(u) \bmod \Lambda \\ q^d+1 & \lambda \equiv \alpha(u) \mod \Lambda \text{ ,} \qquad \text{i.e.} \end{cases}$$
$$T_{\mathfrak{p}} E_u^{(k)}(\Lambda,\alpha) = \sum_{\lambda \in \mathfrak{n}^{-1}\mathfrak{p}^{-1}\Lambda} \lambda^{-k}$$
$$= E_u^{(k)}(\mathfrak{p}^{-1}\Lambda,\alpha') \quad . \qquad \square$$

It is a highly interesting question to describe the action of $\mathcal{H}$ on the modular forms $\Delta_a, \Delta_{\mathfrak{n}}, l_i$ etc., in particular on the canonical form $\Delta$ of (VI 5.14). In principle, one should be able to trail $T_{\mathfrak{p}}$ on the cuspidal expansions. If $A = \mathbb{F}_q[T]$ , the space $S_{q^2-1}(\Gamma)$ is one-dimensional, generated by $\Delta$ which therefore has to be an eigenform. The eigenvalue of $T_{\mathfrak{p}}$ , where $\mathfrak{p} = (f)$ , is $f^{q-1}$ [28], the same eigenvalue as that of $E^{(q-1)}$ !

## 2. Connections with the Classification of Elliptic Curves

Let now $\tilde{M}$ be a modular scheme of conductor $\mathfrak{n}$ , i.e. $\tilde{M}$ lies between $\bar{M}(\mathfrak{n})$ and $\bar{M}(1)$ , and let $\mathfrak{p}$ be a place prime to $\mathfrak{n}$ . Then $T_{\mathfrak{p}}$ induces an endomorphism of the Jacobian $J(\tilde{M}) = J(\tilde{M} \times_A K)$ , also denoted by $T_{\mathfrak{p}}$ . Note that, in general, $\tilde{M} \times K$ is not geometrically irreducible. The Jacobian is defined to be the product of the Jacobians of the components which is provided with a Galois action permuting the components.

The main result of [11] is the interpretation of the l-adic cohomology module $H(\tilde{M}) = H^1(\tilde{M} \times C, \mathbb{Q}_l)$ as a space of automorphic cusp forms in the sense of [40], l denoting a prime number different from $p_\sim$. The representation of the ring of K-endomorphisms of $J(\tilde{M})$ in $H(\tilde{M})$ gives a representation of $\mathcal{H}$ in $H(\tilde{M})$.

We now specialize $\tilde{M}$ to be the modular scheme $\bar{M}_0(\mathfrak{n})$ of Hecke type, i.e.

$$\bar{M}_0(\mathfrak{n}) = K_0(\mathfrak{n}) \backslash \bar{M} , \qquad K_0(\mathfrak{n}) = \left\{ \begin{pmatrix} \underline{a} & \underline{b} \\ \underline{c} & \underline{d} \end{pmatrix} \in GL(2,\hat{A}) \,|\, \underline{c} \equiv 0 \bmod \mathfrak{n} \right\} .$$

Then the following assertions hold:

(2.1) (i) There is a canonical $\mathcal{H}$-stable decomposition

$$J(\bar{M}_0(\mathfrak{n})) \sim J^{old} \times J^{new}$$

("$\sim$" denotes isogeneous) into some factor $J^{new}$ belonging genuinely to the conductor $\mathfrak{n}$, and some part $J^{old}$ derived from proper divisors of $\mathfrak{n}$.

(ii) The image of $\mathcal{H} \otimes \mathbb{Q}$ in $\mathrm{End}(J^{new}) \otimes \mathbb{Q}$ is a semi-simple algebra of degree $\dim(J^{new})$ over $\mathbb{Q}$.

(iii) If U and V are K-irreducible, K-isogeneous $\mathcal{H}$-stable abelian submanifolds of $J^{new}$, the cohomology modules $H^1(U,\mathbb{Q}_l)$ and $H^1(V,\mathbb{Q}_l)$ agree as submodules of $H^1(J^{new},\mathbb{Q}_l)$.

(iv) Each elliptic curve over K having Tate reduction at the place $\infty$ and geometric conductor $\mathfrak{n} \cdot \infty$ occurs up to isogeny in the $\mathcal{H}$-decomposition of $J^{new}$.

The <u>geometric conductor</u> of an elliptic curve E is defined in [52], see also [65]. It is a positive divisor divisible precisely by the places of bad reduction of E. The curve E has <u>Tate reduction</u> at $\infty$ if E degenerates at $\infty$ to a cubic with a double point having rational tangents.

(2.1) (i) follows directly from the construction of the newform associated with an automorphic representation (see [7, 2.4]), combined with [11, 10.3, Thm.2]. Items (ii) and (iii) are implied by the "multiplicity 1" assertion for automorphic representations [40, 11.1.1]. Finally, (iv) results from [40] and [11], combined with a theorem of Grothendieck [9, sect.9+10] on Galois representations over function fields. It is an analogue of the famous conjecture assigned to Taniyama and Weil concerning the parametrization of elliptic curves over $\mathbb{Q}$ by classical modular curves. Using (iv), one may classify elliptic curves over $K$ with the prescribed reduction behavior up to isogeny; at least, one can estimate the number of isogeny classes by

$$\text{(2.2)} \qquad \dim J^{\text{new}} \leqq \dim J(\bar{M}_0(\mathfrak{n})) = h \cdot g(\bar{M}_{\Gamma_0(\mathfrak{n})}) .$$

2.3 <u>Corollary</u>. If $\mathfrak{n}$ is a positive divisor such that $g(\bar{M}_{\Gamma_0(\mathfrak{n})}) = 0$, there are no elliptic curves over $K$ having Tate reduction at $\infty$ and the reduction type prescribed by $\mathfrak{n}$ at the finite places. This assumption is fulfilled if $g(K) = 0$ and $\delta + \deg \mathfrak{n} \leqq 3$ or $g(K) = 1$ with $\delta = 1$ and $\mathfrak{n} = A$.

This assertion for $(g,\delta) = (1,1)$ has also been observed by Stuhler [64].

The operation of $H$ on $H^1(J^{\text{new}})$ may in principle be computed, provided the data $K,\infty,\mathfrak{n}$ are given in a sufficiently explicit form. One uses the connection explained in (V Appendix) between the graph cohomology of $\Gamma_0(\mathfrak{n})\backslash\mathcal{T}$ and $H^1(J(\bar{M}_{\Gamma_0(\mathfrak{n})}))$. In [22], the reader may find tables for the decomposition of $J(\bar{M}_0(\mathfrak{n}))$ in the case $(K,\infty) = (\mathbb{F}_q(T),\infty)$, divisors $\mathfrak{n}$ of degree $\leqq 3$ and constant fields $\mathbb{F}_q$ where $q \leqq 16$. As a result, there are significantly fewer elliptic curves than would be possible by the crude estimation (2.2).

2.4 <u>Remark</u>. A 2-D-module $\phi$ over a finite extension $\mathbb{F}$ of $A/\mathfrak{p}$ ($\mathfrak{p}$ denoting a prime ideal) is called <u>supersingular</u>, if $\phi$ has no $\mathfrak{p}$-division points over $\bar{\mathbb{F}}$, or equivalently, if its endomorphism ring is non-commutative. As for elliptic curves, there is a relation between supersingular D-modules and the geometry of the special fiber of $M_0(\mathfrak{p})$ [8, VI.6]. In the case $(g,\delta) = (0,1)$, this has been carried out in [18,21], where, besides a "mass-formula" for the number of automorphisms, we obtained $g(\bar{M}_{\Gamma_0(\mathfrak{p})}) + 1$ for the number of supersingular isomorphism classes.

## 3. Some Open Questions

Finally, I would like to point out some questions which seem to be particularly interesting. Some of them are related with analogous problems of number theory.

(3.1) For an arithmetic group $\Gamma$ , give a geometric description of the algebra $M(\Gamma) = \oplus M_k(\Gamma)$ of modular forms, i.e. a presentation by generators and relations. Especially, which relations hold between the Eisenstein series? Where do their zeroes lie? How does the Hecke algebra act on $M_k(\Gamma)$ ?

(3.2) Investigate the properties of the expansion coefficients of the modular forms $\Delta_n$ resp. $\Delta$ ! What about congruence properties and relations with Galois representations?

(3.3) Assuming $(i,\delta) = 1$ , the place $\infty$ remains prime in $K_i = K \cdot \mathbb{F}_{q^i}$ . On the one hand, D-modules of rank r for $(K_i,\infty)$ may be considered as D-modules of rank $ir$ with "complex multiplication" for $(K,\infty)$(in (VII sect. 2), we have made use of this phenomenon); on the other hand, one may, for instance, compare the modular schemes $M^2$ for $(K,\infty)$ and for $(K_i,\infty)$ and describe the "base change" of automorphic forms as defined in [48]. The details of the relations between the different modular schemes need to be carried out.

(3.4) The space of differentials $H^0(\bar{M}_\Gamma,\mathfrak{D})$ on a modular curve $\bar{M}_\Gamma$ and $\underline{H}_!^1(\mathbb{C},\mathbb{Q})^\Gamma$ are related: Both modules have equal dimension over C resp. over $\mathbb{Q}$ , and there are compatible actions of those Hecke operators that respect the component $\bar{M}_\Gamma$ . Are there canonical bases of the vector spaces in question, and a bijection which can be described by means of the building map? While $\underline{H}_!^1$ has a basis consisting of eigenvectors, the situation for $H^0(\mathfrak{D})$ is mysterious.

(3.5) Compute the order of the group of cuspidal divisor classes of degree zero on $\bar{M}_{\Gamma'}$ , for example for the various groups $\Gamma' = \Gamma(n) \subset \Gamma_1(n) \subset \Gamma_0(n) \subset \Gamma = GL(2,A)$ ! The examples (VI 5.17) show that already in case $g(\bar{M}_\Gamma) = 0$ , it is not easy to describe the group of modular functions whose divisors are supported by the cusps. The order of such groups would be of interest in view of the work of Ribet,

Wiles, and Mazur [50,51,55,70], where groups of this type play an important part.

(3.6) What is the rule of splitting of $J^{new}$ with respect to $\mathbb{H}$ ? This seems to be a very deep question: As far as I know, already in the case of cusp forms for the classical modular group $SL(2,\mathbb{Z})$ , there is no general answer to the question of splitting!

Index


| | |
|---|---|
| admissible divisor | 5 |
| arithmetic subgroup | 44 |
| Bruhat-Tits building | 41 |
| building map | 41 |
| characteristic of an A-algebra | 2 |
| cusp | 44 |
| distribution property | 16 |
| divisor of a modular form | 85 |
| Drinfeld module, D-module, r-D-module, isogeny, isomorphism, rank, division point | 3,4 |
| Eisenstein series | 14,47 |
| elliptic point | 50 |
| exponential function of a lattice | 6 |
| Hecke algebra, operator, | 94 |
| congruence subgroup | 90 |
| Hilbert class field | 28 |
| Hurwitz formula | 87 |
| ideal class group, | 18 |
| narrow | 28 |
| imaginary absolute value | 40 |
| invariant of a 1-lattice | 30,38 |
| lattice in $C$ | 6 |
| in $K^r$ | 10 |
| leading coefficient of an additive polynomial | 3 |
| level, structure of | 5 |
| modular form for $\Gamma$, | 47 |
| of level $n$, | 49 |
| algebraic | 80 |
| monic | 25,27 |
| normalizing field | 28 |
| normalized D-module | 27 |
| p'-torsion free | 44 |
| Riemann-Roch theorem | 17 |
| ramified (wildly, tamely) | 87 |
| Serre duality | 17 |
| sgn-normalized D-module | 27 |
| sign function, twisted | 27 |
| similar lattices | 6 |
| Weierstraß gap, non-gap | 21 |
| Z-function, zeta function, partial | 18,19 |

List of Symbols


| Page | Symbols |
|---|---|
| 1 | $q$, $\mathbb{F}_q$, $K$, $g$, $\infty$, $\delta$, $A$, $K_\infty$, $O_\infty$, $\mathfrak{k}$, $\pi$, $\lvert x\rvert$, $\lvert\mathfrak{a}\rvert$, $\mathfrak{a}_N$, $\mathfrak{a} > 1$, $\mathbb{A}$, $\mathbb{A}_f$, $I$, $I_f$, $E$, $E_f$, $E_\infty$, $E(\mathfrak{a})$ |
| 2 | $\hat{A}$, $\mathrm{End}_L(G_a)$, $\tau_p$ |
| 3 | $\tau$, $l_i(f)$, $D(f)$, $l(f)$ |
| 4 | $D(\phi,a)$, $D(\phi,\mathfrak{n})$ |
| 5 | $r$-$\mathfrak{DM}$, $r$-$\mathfrak{DM}(\mathfrak{n})$, $M^r(\mathfrak{n})$, $C$ |
| 6 | $e_\Lambda$ |
| 7 | $\phi^\Lambda$ |
| 8 | $M^r$, $G(\hat{A},\mathfrak{n})$ |
| 9 | $M_K^r$, $\bar{M}^2(\mathfrak{n})$ |
| 10 | $\bar{M}^2$, $\bar{M}_K^2$, $Y(\underline{g})$ |
| 11 | $\mathfrak{P}_A^r$ |
| 12 | $\Omega^r$, $\Gamma_{\underline{x}}$ |
| 14 | $E^{(k)}(\Lambda)$ |
| 15 | $l_i(a,\Lambda)$, $\mu(\Lambda,\Lambda')$ |
| 16 | $\mathfrak{n}*\phi^\Lambda$ |
| 17 | $J$ |
| 18 | $\zeta_K(s)$, $Z_K(S)$, $P(S)$, $h$, $\zeta_A$, $Z_A$, Pic $A$, $\mathfrak{a} \sim \mathfrak{b}$, $(\mathfrak{a})$, $\zeta_{(\mathfrak{a})}$, $Z_{(\mathfrak{a})}$ |
| 19 | $\zeta_{a,\mathfrak{n}}$, $Z_{a,\mathfrak{n}}$ |
| 20 | $i^*$, $i_*$, $m(\mathfrak{n})$ |
| 21 | $r(a,\mathfrak{n})$, $w(a,\mathfrak{n})$ |
| 22 | $Q_i$ |
| 27 | sgn |
| 28 | $H$, $\tilde{H}$, $H(\mathfrak{n})$, $\tilde{H}(\mathfrak{n})$, $B$, $\tilde{B}$, $B(\mathfrak{n})$, $\tilde{B}(\mathfrak{n})$, $\widetilde{\text{Pic}}\ A$ |
| 30 | $\xi(\Lambda)$, $\varepsilon_{(\mathfrak{a})}$ |
| 31 | $\varepsilon_{u,\mathfrak{a}}$ |
| 32 | $U_{u,\mathfrak{a}}$ |
| 36 | $\mathrm{sgn}(f,\mathfrak{a})$ |

38 $\Lambda^{(\mathfrak{a})}$, $\rho^{(\mathfrak{a})}$, $\Theta(\mathfrak{a},\mathfrak{b})$

39 $e_{(\mathfrak{b})}$

40 $\Omega$, $|z|_i$

41 $\mathfrak{T}$, $\mathfrak{T}(\mathbb{Z})$, $\mathfrak{T}(\mathbb{R})$, $\lambda$, $\mathfrak{T}(\mathbb{Q})$

42 $\mathfrak{T}(r)$, $B(z,r)$, $B(\infty,r)$

44 $\lambda_\Gamma$, $M_\Gamma$, $\bar{M}_\Gamma$

45 $Sp(\Gamma)$, $U(Y,s)$, $V(Y,s)$

46 $t(\nu,\Gamma)$

47 $f_{[\gamma]_k}$, $M_k(\Gamma)$, $S_k(\Gamma)$, $Y_\omega$, $E^{(k)}(\omega)$

48 $\Delta_a$, $\Delta_\mathfrak{a}$

49 $e_u$, $E_u^{(k)}$

50 $Ell(\Gamma)$

59 $t_\mathfrak{g}$, $R_\mathfrak{n}^{(\mathfrak{g})}$

72 $s(\mathfrak{a},\mathfrak{b})$

76 $\Delta$

79 $C(f)_0$, $C(M_\Gamma)$, $f_u(\omega)$

80 $K(M(\mathfrak{n}))$

85 $[D]$

89 $h(\mathfrak{n})$

94 $[\Lambda':\Lambda]$, $T_\mathfrak{m}$, $S_\mathfrak{m}$, $\mathfrak{H}$

Bibliography


[1] E. Artin - J. Tate: Class Field Theory. New York, Amsterdam: Benjamin 1967

[2] S. Bosch - U. Güntzer - R. Remmert: Non-Archimedean Analysis. Berlin, Heidelberg, New York, Tokyo: Springer 1984

[3] N. Bourbaki: Algèbre commutative. Paris: Hermann 1969

[4] L. Carlitz: On certain functions connected with polynomials in a Galois field. Duke Math. J.1, 137-168, 1935

[5] L. Carlitz: An analogue of the von Staudt-Clausen theorem. Duke Math. J.3, 503-517, 1937

[6] L. Carlitz: An analogue of the von Staudt-Clausen theorem. Duke Math. J.7, 62-67, 1940

[7] P. Deligne: Formes modulaires et représentations de GL(2). Lecture Notes in Mathematics 349. Berlin, Heidelberg, New York: Springer 1973

[8] P. Deligne - M. Rapoport: Les schémas de modules de courbes elliptiques. Lecture Notes in Mathematics 349. Berlin, Heidelberg, New York: Springer 1973

[9] P. Deligne: Les constantes des équations fonctionnelles des fonctions L. Lecture Notes in Mathematics 349. Berlin, Heidelberg, New York: Springer 1973

[10] P. Deligne - D. Husemoller: Survey of Drinfeld Modules. Unpublished Notes. Bures-sur-Yvette 1977

[11] V.G. Drinfeld: Elliptic Modules (Russian). Math. Sbornik 94 594-627,1974.English Translation: Math. USSR-Sbornik 23 No.4, 561-592, 1976

[12] V.G. Drinfeld: Elliptic Modules II. Math. USSR-Sbornik 31 No.2, 159-170, 1977

[13] J. Fresnel - M. van der Put: Géometrie Analytique Rigide et Applications. Progress in Mathematics 18. Boston, Basel, Stuttgart: Birkhäuser 1981

[14] S. Galovich - M. Rosen: The class number of cyclotomic function fields. J.Number Theory 13, 363-375, 1981

[15] S. Galovich - M. Rosen: Units and class groups in cyclotomic function fields. J.Number Theory 14, 156-184, 1982

[16] S. Galovich - M. Rosen: Distributions on Rational Function Fields. Math. Annalen 256, 549-560, 1981

[17] E. Gekeler: Drinfeld-Moduln und modulare Formen über rationalen Funktionenkörpern. Bonner Math. Schriften 119, 1980

[18] E. Gekeler: Zur Arithmetik von Drinfeld-Moduln. Math. Annalen 262, 167-182, 1983

[19] E. Gekeler: A Product Expansion for the Discriminant Function of Drinfeld Modules of Rank Two. J.Number Theory 21, 135-140, 1985

[20] E. Gekeler: Modulare Einheiten für Funktionenkörper. J.reine angew. Math. 348, 94-115, 1984

[21] E. Gekeler: Über Drinfeld'sche Modulkurven von Hecke-Typ. Comp. Math.57, 219-236, 1986

[22] E. Gekeler: Automorphe Formen über $\mathbb{F}_q(T)$ mit kleinem Führer. Abh. Math. Sem. Univ. Hamburg 55, 111-146, 1985

[23] E. Gekeler: Le genre des courbes modulaires de Drinfeld. C.R. Acad. Sc. Paris, t. 300, Série I no.19, 1985

[24] L. Gerritzen - M. van der Put: Schottky Groups and Mumford Curves. Lecture Notes in Mathematics 817. Berlin, Heidelberg, New York: Springer 1980

[25] O. Goldman - N. Iwahori: The space of p-adic norms. Acta Math. 109, 137-177, 1963

[26] D. Goss: Von Staudt for $\mathbb{F}_q[T]$ .Duke Math. J.45, 885-910, 1978

[27] D. Goss: $\pi$-adic Eisenstein Series for Function Fields. Comp. Math.41, 3-38, 1980

[28] D. Goss: Modular Forms for $\mathbb{F}_r[T]$ . J.reine angew. Math.317, 16-39, 1980

[29] D. Goss: Kummer and Herbrand criterion in the theory of function fields. Duke Math. J.49, 377-384, 1982

[30] D. Goss: On a new type of L-function for algebraic curves over finite fields. Pac. J. Math.105, 143-181, 1983

[31] G. Harder: Halbeinfache Gruppenschemata über Dedekindringen. Inv.Math.4, 165-191, 1967

[32] G. Harder: Minkowskische Reduktionstheorie über Funktionenkörpern. Inv.Math.7, 33-54, 1969

[33] G. Harder: Eine Bemerkung zu einer Arbeit von P.E. Newstaed. J.reine angew. Math.242, 16-25, 1970

[34] G. Harder: Chevalley Groups over Function Fields and Automorphic Forms. Ann. Math.100 No.2, 249-306, 1974

[35] D. Hayes: Explicit class field theory for rational function fields. Trans. Amer. Math. Soc. 189, 77-91, 1974

[36] D. Hayes: Explicit class field theory in global function fields. Studies in Algebra and Number Theory. G.C.Rota ed. New York: Academic Press 1979

[37] D. Hayes: Analytic class number formulas in global function fields. Inv. Math.65, 49-69, 1981

[38] D. Hayes: Elliptic units in function fields. Proc. Conf. on Modern Developments Related to Fermat's Last Theorem. D. Goldfeld ed. Boston, Basel, Stuttgart: Birkhäuser 1982

[39] D. Hayes: Stickelberger Elements in Function Fields. Comp. Math. 55, 209-239, 1985

[40] H. Jacquet - R.P. Langlands: Automorphic Forms on GL(2) . Lecture Notes in Mathematics 114. Berlin, Heidelberg, New York: Springer 1970

[41] N. Katz: P-adic properties of modular schemes and modular forms. Lecture Notes in Mathematics 350. Berlin, Heidelberg, New York: Springer 1973

[42] N. Katz - B. Mazur: Arithmetic Moduli of Elliptic Curves. Ann. Math. Studies 108. Princeton: Princeton University Press 1985

[43] R. Kiehl: Der Endlichkeitssatz für eigentliche Abbildungen in der nichtarchimedischen Funktionentheorie. Inv. Math. 2, 191-214, 1967

[44] R. Kiehl: Theorem A und Theorem B in der nichtarchimedischen Funktionentheorie. Inv. Math. 2, 256-273, 1967

[45] D. Kubert - S. Lang: Modular Units. Berlin, Heidelberg, New York: Springer 1981

[46] E.E. Kummer: Mémoire sur la théorie des nombres complexes composés de racines de l'unité et de nombres entiers. J.de Math. 16, 377 -498,1851

[47] S. Lang: Elliptic Functions. Reading: Addison-Wesley 1973

[48] R.P. Langlands: Base Change for GL(2) . Ann. Math. Studies 96. Princeton: Princeton University Press 1981

[49] B.H. Matzat: Über Weierstraßpunkte von Fermatkörpern. Dissertation Karlsruhe 1972

[50] B. Mazur: Modular curves and the Eisenstein ideal. Publ. Math. IHES 47, 1977

[51] B. Mazur - A. Wiles: Class fields of abelian extensions of $\mathbb{Q}$ . Inv. Math. 76, 179-330,1984

[52] A. Ogg: Elliptic curves and wild ramification. Amer. J. Math. 89, 1-21, 1967

[53] O. Ore: On a special class of polynomials. Trans. Amer. Math. Soc. 35, 559-584, 1933

[54] W. Radtke: Diskontinuierliche Gruppen im Funktionenkörperfall. Dissertation Bochum 1984

[55] K. Ribet: A modular construction of unramified p-extensions of $\mathbb{Q}(\mu_p)$ . Inv. Math 34, 151-162, 1976

[56] G. Robert: Unités elliptiques. Bull. Soc. Math. France mémoire no. 36, 1973

[57] J.P. Serre: Groupes algébriques et corps de classes. Paris: Hermann 1959

[58] J.P. Serre: Corps locaux. Paris: Hermann 1968

[59] J.P. Serre: Cours d'arithmétique. Paris: Presses universitaires de France 1970

[60] J.P. Serre: Cohomologie des groupes discrets. Ann. Math. Studies 70. Princeton: Princeton University Press 1971

[61] J.P. Serre: Arbres, Amalgames, $SL_2$ . Astérisque 46, 1977

[62] G. Shimura: Introduction to the arithmetic theory of automorphic forms. Publ. Math. Soc. Japan 11, Tokyo, Princeton 1971

[63] W. Sinnott: On the Stickelberger ideal and the circular units of a cyclotomic field. Ann. Math. 108, 107-134, 1978

[64] U. Stuhler: Über die Faktorkommutatorgruppe der Gruppen $SL_2(O)$ im Funktionenkörperfall. Arch. Math. 42, 314-324, 1984

[65] J. Tate: Algorithm for determing the Type of a Singular Fiber in an Elliptic Pencil. Lecture Notes in Mathematics 476. Berlin, Heidelberg, New York: Springer 1975

[66] J. Tate: Les conjectures de Stark sur les Fonctions L d'Artin en $s = 0$ . Progress in Mathematics 47. Boston, Basel, Stuttgart: Birkhäuser 1984

[67] A. Weil: Sur les courbes algébriques et les variétés qui s'en déduisent. Paris: Hermann 1948

[68] A. Weil: Basic Number Theory. Berlin, Heidelberg, New York: Springer 1967

[69] A. Weil: Dirichlet Series and Automorphic Forms. Lecture Notes in Mathematics 189. Berlin, Heidelberg, New York: Springer 1971

[70] A. Wiles: Modular curves and the class group of $\mathbb{Q}(\zeta_p)$ . Inv. Math. 58, 1-35, 1980

[71] J. Yu: A Six Exponentials Theorem in Finite Characteristic. Math. Annalen 272, 91-98, 1985

[72] J. Yu: Transcendence and Drinfeld Modules. Inv. Math. 83, 507-517, 1986

[73] E. Gekeler: Compactification of Drinfeld Modular Schemes. In Preparation